21世纪高等院校网络工程规划教材
21st Century University Planned Textbooks of Network Engineering

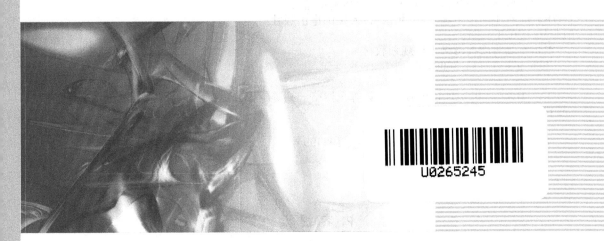

计算机网络实践教程
——基于GNS3网络模拟器（CISCO技术）

Computer Network Practice

王文彦 主编
傅秀芬 黄益民 陈鑫杰 副主编

人民邮电出版社
北京

图书在版编目（CIP）数据

计算机网络实践教程：基于GNS3网络模拟器：CISCO技术 / 王文彦主编. -- 北京：人民邮电出版社，2014.10（2023.1重印）
 21世纪高等院校网络工程规划教材
 ISBN 978-7-115-36692-4

Ⅰ. ①计… Ⅱ. ①王… Ⅲ. ①计算机网络－高等学校－教材 Ⅳ. ①TP393

中国版本图书馆CIP数据核字(2014)第199499号

内 容 提 要

本书系统地介绍了计算机网络技术的基本原理和基于 GNS3 网络模拟器的实用操作技术。

本书共分为 10 章，主要内容包括：GNS3 安装与使用，设备管理，TCP/IP 协议栈，路由技术，交换技术，广域网技术，安全策略，高级安全，企业项目实战，综合测试等。

本书可作为普通高等院校、高职高专院校网络技术的教材使用，也可作为社会培训机构的培训用书。

◆ 主　编　王文彦
　　副 主 编　傅秀芬　黄益民　陈鑫杰
　　责任编辑　许金霞
　　责任印制　彭志环　杨林杰

◆ 人民邮电出版社出版发行　北京市丰台区成寿寺路 11 号
　　邮编　100164　电子邮件　315@ptpress.com.cn
　　网址　http://www.ptpress.com.cn
　　北京天宇星印刷厂印刷

◆ 开本：787×1092　1/16
　　印张：19.25　　　　　　2014 年 10 月第 1 版
　　字数：543 千字　　　　 2023 年 1 月北京第 12 次印刷

定价：42.00 元

读者服务热线：(010)81055256　印装质量热线：(010)81055316
反盗版热线：(010)81055315

前　言

互联网的发展为我们带来了巨大的便利以及生活与学习上的改变，目前全国许多高校开设了计算机网络相关的专业和课程，同时社会上也有许多计算机网络爱好者，想了解和学习其技术。计算机网络技术是一门实践性很强的课程，要想很好的掌握它，需要在学习一定的理论基础之后，通过大量的实践操作，才能够将理论结合实际，取得良好的学习效果。

目前，图书市场上关于计算机网络技术的教材有好多，大多数侧重理论的讲解，理论与实践结合的比较少。关于实验和实践方面的教材基本上都需要大量的网络硬件设备等支持，对于开设课程的学校和想要掌握实际操作的学生来说，是无法很好的实现的。尤其是许多学校缺少或仅有少量的网络设备，或者拥有的设备厂商、型号等与采用的教材不匹配，这些都给我们的学校、老师和学生带来许多的困扰。

另外，在各高校的不断扩招，以及全国600所本科院校转型发展的环境下，许多学校都面临着学生多，实验设备和经费不足，实验教师少，实验课时少，实验内容大大缩水，实验内容与社会企业需求脱节等情况。

针对上述情况，作者根据多年的网络工程经验和教学实践经验编写了本书。本书在编写中注重原理与实践紧密结合，采用目前计算机网络行业最好的开源模拟器来实现网络实验、网络配置的完全仿真，为学校节省大量的设备费用，为教师和学生提供全真的操作界面和环境，同时针对社会企业需求，加强学生的实践能力。本书的电子版已在广东工业大学校内试用两年，获得师生的广泛好评，已有很多学生通过学习，提高了网络实践能力，进而得到了社会知名IT企业的认可。

【本书特色】

1. 采用的网络实验技术为全球最主流的思科网络技术，这些技术也同样适用于国内比较流行的华为、中兴、锐捷等厂商技术。

2. 采用开源的GNS3软件全真模拟实验，这些操作与在思科设备上实际操作完成相同。

3. 每个网络实验配有原理和操作步骤详细的讲解，可减少学生的陌生感和跳跃感，适合新手和熟手等不同层次读者的学习。各学校可根据本校相关课程的内容和课时，选取部分实验内容进行实践。

4. 在表现形式方面，实验中的原理解释让学生将原理和技术融合。实验随手记可帮助学生在预习时记录疑惑，方便在实验课上咨询老师，也可用来记录实验中的数据和结果。书页的右栏术语和名词解释像网页中的超链接一样帮助学生学习，避免学生再查找相关知识与解释，防止学生顺便被"拐走"到其他网页，回不到网络实验中来。

5. 本书网络实验较多，各学校可根据本校相关课程的内容和课时，选取部分实验内容进行实践操作。其他内容学生可在课外时间,安装本书中的模拟软件进行全部的实践操作。

6. 本书附综合测试,以加强对各章基础知识的理解与掌握。并配有实操部分的实验报告。

7. 本书中的网络技术思维导图,可便于学生融会贯通网络体系结构与不同技术之间的联系。

【本书结构】

本书每章中包括要点说明和所有实验的思维导图,便于学生理清关系,全面掌握,也便于学生在做完实验后总结和梳理。本书还设计了适量的习题和实操检验,主要是针对每章的重点、难点进行训练。

全书共分为 10 章,主要内容包括:GNS3 安装与使用,设备管理,TCP/IP 协议栈,路由技术,交换技术,广域网技术,安全策略,高级安全,企业项目实战,综合测试,附录等。

本书免费附赠习题答案和实验报告,教师可登录人民邮电出版社教学资源与服务网(www.ptpedu.com.cn)下载。

【课时建议】

采用本教程作为计算机网络技术理论课程的实验指导书,在实验课时为 8 学时的情况下,建议由学生先在自己的电脑上自学完成本书的第 1 章,先学会 GNS3 的安装与使用。利用 2 学时完成设备管理实验,2 学时完成 TCP/IP 中的抓包分析实验,2 学时完成路由技术中的静态路由和 RIP 实验,2 学时完成交换技术中的 VLAN 和单臂路由实验。

在实验课时为 16 学时的情况下,建议在上面的实验基础上增加一些实验,包括 2 学时 OSPF 实验,2 学时 EIGRP 实验,2 学时 STP 实验,2 学时三层交换实验。

在实验课时为 24 学时的情况下,建议在上面的实验基础上再增加部分实验,包括 2 学时 DHCP 实验,2 学时 PPP 实验,2 学时 ACL 实验,2 学时 NAT 实验。

如采用本书作为计算机网络实训教材,实训时间为 2~3 周。在学生基本掌握计算机网络理论的基础上,可基本完成所有实验。如果实训进度较慢,很多学生也可利用课余时间完成剩余的实验。

建议各学校可成立网络工程兴趣小组,组织学生来学习和实践本书中的所有实验,这样将会大大提升学生的学习兴趣和动手能力,使得学生在学习和实践中与社会需求接轨。

本书由王文彦担任主编,傅秀芬、黄益民、陈鑫杰任副主编。参与编写的还有滕少华、林穗、刘东宁、何翠红、孙为军、梁路、申建芳、丁国芳、彭重嘉、陈靖宇等。程良伦教授担任主审,并提出许多宝贵建议。

本书中的所有实验由深圳拼客科技公司采用全真机实验环境验证通过,在此对拼客科技公司的鼎力支持表示感谢。本书在编写过程中也得到了思科系统网络技术有限公司的大力支持,在此表示深深的谢意。

由于作者水平所限和时间仓促,书中难免出现不足,望读者给予指正。作者邮箱:fwwy@gdut.edu.cn

王文彦

2014 年 7 月

目 录

第 1 章　GNS3 安装与使用 ········ 1
1.1　GNS3 简要介绍 ············· 2
1.2　GNS3 安装调试 ············· 3
1.3　GNS3 拓扑创建 ············· 7

第 2 章　设备管理 ············· 12
2.1　操作模式 ··················· 13
2.2　初始管理 ··················· 16
2.3　时间管理 ··················· 18
2.4　接口管理 ··················· 20
2.5　配置管理 ··················· 24
2.6　密码管理 ··················· 28
2.7　IOS 管理 ··················· 32

第 3 章　TCP/IP 协议栈 ········ 35
3.1　ARP ························· 36
3.2　IP ··························· 40
3.3　ICMP ······················· 44
3.4　UDP&DHCP ··············· 50
3.5　TCP&Telnet ··············· 55

第 4 章　路由技术 ············ 61
4.1　静态路由 ··················· 62
4.2　默认路由 ··················· 66
4.3　浮动路由 ··················· 70
4.4　RIPv1 基本配置 ··········· 76
4.5　RIPv2 基本配置 ··········· 81
4.6　RIPv2 路由汇总 ··········· 86
4.7　EIGRP 基本配置 ··········· 92
4.8　EIGRP 路由汇总 ··········· 99

4.9　OSPF 基本配置 ············ 105
4.10　OSPF 多区域配置 ········ 111
4.11　OSPF 路由汇总 ··········· 116

第 5 章　交换技术 ············ 122
5.1　VLAN 基本配置 ············ 123
5.2　VLAN 进阶配置 ············ 128
5.3　Trunk 基本配置 ············ 132
5.4　Trunk 进阶配置 ············ 138
5.5　DTP 基本配置 ············· 145
5.6　VTP 基本配置 ············· 148
5.7　STP 基本配置 ············· 156
5.8　STP 进阶配置 ············· 161
5.9　PVST 基本配置 ············ 164
5.10　单臂路由 ··················· 168
5.11　三层交换机 ················· 171
5.12　DHCP 基本配置 ··········· 175
5.13　L2 Etherchannel 基本配置···179

第 6 章　广域网技术 ·········· 183
6.1　HDLC 基本配置 ············ 184
6.2　PPP 基本配置 ············· 187
6.3　PPP PAP 认证 ············· 190
6.4　PPP CHAP 认证 ··········· 193
6.5　PPP Multilink ············· 195
6.6　Frame-Relay 基本配置 ····199
6.7　Frame-Relay&Static Route ···206
6.8　Frame-Relay&RIPv2 ······211
6.9　Frame-Relay&EIGRP ······217

第 7 章　安全策略 ················ 223

7.1 编号标准 ACL ············· 224
7.2 编号拓展 ACL ············· 228
7.3 命名 ACL ················· 231
7.4 时间 ACL ················· 236
7.5 动态 NAT ················· 240
7.6 端口 NAT ················· 245
7.7 静态 NAT ················· 250

第 8 章　高级安全 ················ 254

8.1 防火墙的基本操作 ········· 255
8.2 防火墙的进阶操作 ········· 267
8.3 防火墙的高级应用 ········· 273

第 9 章　企业项目实战 ············ 279

9.1 项目实战一 ··············· 280
9.2 项目实战二 ··············· 282

第 10 章　综合测试 ··············· 285

10.1 设备管理测试 ············ 286
10.2 TCP/IP 协议栈测试 ······· 288
10.3 路由技术测试 ············ 291
10.4 交换技术测试 ············ 293
10.5 广域网技术测试 ·········· 295
10.6 安全策略测试 ············ 297
10.7 高级安全测试 ············ 299

附　录　术语索引 ················ 301

第 1 章　GNS3 安装与使用

本章主要学习网络技术模拟器 GNS3 的安装与使用，除此之外，我们还将了解抓包软件 Wireshark、终端登录软件 SecureCRT 的使用和功能。通过这几个软件的使用，可以利于我们后续实验环境的搭建，减少搭建真机的成本和时间。以下为本章导航图。

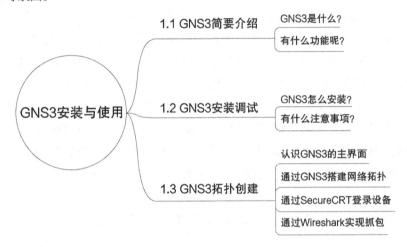

1.1 GNS3 简要介绍

GNS3 软件是一个图形化的网络模拟器，可以模拟复杂的网络，例如它能够完整地模拟整个校园网络或企业网络。不仅如此，GNS3 还是一个跨平台的软件，可以同时在 Windows、Linux 或者 Mac OS X 上进行部署。

GNS3 是由多个组件集合而成的，包含了 Dynamips、Qemu、Wireshark 等程序。下面我们对这些组件做一个基本的介绍。Dynamips 是一个基于虚拟化技术的模拟器 (emulator)，本身就能够模拟路由器和交换机，但是 Dynamips 是命令行界面的，对于新手而言，还是有很大的挑战，所以 GNS3 在 Dynamips 的基础上，加入了一个非常友好的图形化操作界面；Qemu 可以允许我们在 GNS3 上面模拟防火墙、入侵检测系统、Juniper 路由器等；Wireshark 则可以让我们抓取网络设备之间的数据流并进行底层分析。

目前，在众多网络模拟器中，GNS3 是功能最全、用户体验最佳的模拟器。由于 GNS3 模拟的是路由器和交换机等网络设备，所有需要调用网络操作系统镜像如思科的 IOS 系统镜像，而且我们调用的是真实的网络操作系统，所以当我们通过 GNS3 搭建模拟环境时，敲打的命令和输出的内容与真机没有任何区别。而其他模拟器会经常出现命令不支持或者调试失败等情况。不仅如此，GNS3 还是开源免费的。

另外一个网络模拟器是由思科公司官方出品的 Cisco Packet Tracer，它是思科公司针对其 CCNA 认证所推出的。相比 GNS3 而言，Cisco Packet Tracer 的主要功能侧重于路由和交换技术，而且没法对真正的数据进行抓包，没法与真实网络进行桥接和融合。对于希望深入研究网络的朋友而言，GNS3 的拓展性要远远高于 Cisco Packet Tracer。

GNS 和 Cisco Packet Tracer 的功能特性对比如下表所示。

	GNS3	Cisco PT
是否支持路由交换功能	支持	支持
是否支持网络安全技术	支持，能实现 IPsec/SSL VPN、防火墙、入侵检测	不支持
是否支持数据抓包功能	支持，默认调用 Wireshark 进行抓包	不支持
是否支持网络桥接	支持，能桥接到真实网络环境，也可以桥接到其他虚拟平台如 VMware 和 VirtualBox，实现跨平台模拟	不支持
是否支持其他厂商	支持，除了支持 Cisco 设备，还支持 Juniper 设备	不支持

总而言之，GNS3 是我们网络工程师入门需要掌握的首要工具，也是帮助我们考取思科网络工程师认证 CCNA/CCNP/CCIE 的好帮手，同时，它还是我们后续部署网络工程项目的首选模拟平台。本书对 GNS3 的安装使用和功能介绍都在 Windows 系统环境下进行讲解。

【GNS3】
GNS3 即 Graphical Network Simulator，图形化网络模拟器。Jeremy Grossmann 是 GNS3 项目的发起人。

【命名行界面】
类似 DOS 界面，对设备的操控需要采用命令输入。

【图形化操作界面】
类似 Windows 操作系统，对设备的操控采用图形界面，使其更显人性化。

【CCNA】
思科认证网络助理工程师

【CCNP】
思科认证网络专业工程师

【CCIE】
思科认证网络专家

1.2 GNS3 安装调试

本节主要介绍 GNS3 在 Windows 系统下的安装和使用步骤。

1.2.1 下载 GNS3

首先我们需要从 GNS3 官网（http://www.gns3.net/）下载 GNS3。最新版本为 GNS3 v0.8.6，建议下载 all-in-one 版本，如图 1-1 所示。

【all-in-one】此版本可以兼容 32 和 64 位系统，以及所有的 GNS3 插件。

图 1-1　GNS3 软件下载

1.2.2 安装 GNS3

①找到从官网下载的 GNS3 安装文件，双击进行安装。如图 1-2 所示。

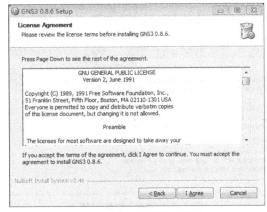

图 1-2　GNS3 软件安装 1

GNS3 安装向导打开后，剩下的事情就是单击"Next"（下一步）或者"Iagree"（同意）或者"Finish"（完成）按钮，其他选项默认，如图 1-3 所示。安装过程中，会提示安装 WINPCAP、Wireshark 等程序，如果电脑以前安装过则不用重复安装。

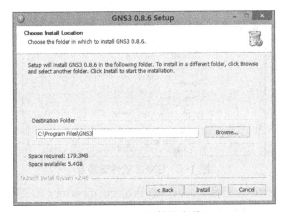

图 1-3　GNS3 软件安装 2

②选择 GNS3 安装目录，可以自行修改默认路径，如图 1-4 所示。

第 1 章　GNS3 安装与使用

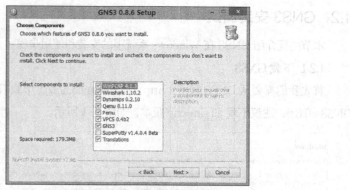

图 1-4　GNS3 软件安装 3

③安装完成后启动 GNS3，屏幕上会出现设置向导提示，提示用户需要三步设置便能够正常使用，包括【设置 IOS 映像文件路径】【检查 Dynamips 工作是否正常】【设定 IOS 对应的 IDLE-PC 值】，后续有详细介绍。如图 1-5 所示。

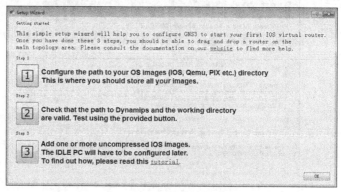

图 1-5　GNS3 软件安装 4

1.2.3　配置 GNS3

①首先进入 Edit（编辑）菜单，选择 Preferences（首选项），把 Language 改为中文。之后重启 GNS3，便可以进入中文界面，如图 1-6 所示。

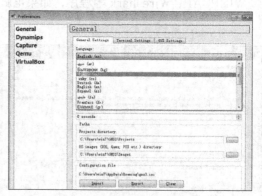

图 1-6　GNS3 软件安装 5

②接下来是 Paths（路径），第一个 Projects directory 为工程目录，此目录用来存放拓扑文件和配置信息；第二个为 OS images directory（OS 镜像目录），此目录用来存放各种系统镜像文件。这两个目录都可以自己设定，也可以选用系统默认的，这里我们使用系统默认的目录，如图 1-7 所示。

1.2　GNS3 安装调试

图 1-7　GNS3 软件安装 6

③单击左侧的 Dynamips，再单击下面的 Test Setting 来测试 Dynamips 是否正常，如正常则出现如图 1-8 所示的提示，如失败则需要检查上面的 Executable path to Dynamips 路径是否正确。

图 1-8　GNS3 软件安装 7

④在 Capture 中勾选 ☑当抓包时自动开始该命令，调试好后，我们单击"OK"按钮，会提示是否创建 project 和 images 目录，我们单击 Yes，如图 1-9 所示。

图 1-9　GNS3 软件安装 8

1.2.4　配置 IOS 文件路径

我们前面提到，GNS3 需要使用 Cisco IOS 镜像文件来模拟路由器和交换机。目前支持的 IOS 平台包括 CISCO7200、3600 系列 (3620、3640 和 3660)、3700 系列 (3725、3745)、2600 系列 (2610 到 2650XM、2691)IOS。

单击主界面的 Edit 菜单，选择 IOS image and hypervisors，系统会弹出对话框。单击 IOS Images 选项框后面的 按钮，默认打开的目录是上面设置的 images 路径，然后选定此目录的 IOS 文件选择打开。该 IOS 文件就会出现在上面的镜像文件信息框中，如图 1-10 所示。

图 1-10　GNS3 软件安装 9

当然，我们也可以修改默认的镜像文件目录，只要保证目录是全英文路径即可。GNS3 会自动识别此 IOS 文件的平台和型号，当 IOS 系统开始运行时，一般会消耗掉我们电脑非常高的 CPU 使用率，所以需要降低 GNS3 的资源消耗，提高 GNS3 的运行效率。这就需要设置 IDLE-PC 值。IDLE-PC 值是 GNS3 用于计算系统消耗的参数，该值直接影响 GNS3 对我们电脑的 CPU 资源占用，好的 IDLE-PC 值可以将 CPU 占用降低到 10% 以下。此处，单击 Auto calculation 按钮，它会自动计算一个 IDLE-PC 值，此过程可能会导致系统卡顿，请耐心等待，直到出现如图 1-11 所示的界面，单击"保存"按钮保存当前的设置，然后单击"关闭"按钮返回 GNS3 主界面。

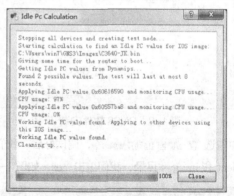

图 1-11　GNS3 软件安装 10

1.3 GNS3 拓扑创建

1.3.1 GNS3 的界面介绍

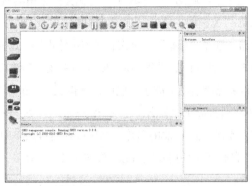

图 1-12　GNS3 界面介绍

GNS3 窗口默认分为四个面板，如图 1-12 所示。左侧的面板列出了可用的节点类型（node），在这里我们可以看到各种路由器、防火墙、以太网交换机等图标，当需要搭建拓扑时，便可从这里拖曳出设备。右侧面板提供 Captures（抓包信息）和 Topology Summary（拓扑汇总概要信息）。中间部分包括两个面板，上面的面板是我们的主要工作区，用于图形化显示拓扑结构。下面的面板称为 Console 面板，显示 Dynagen 的工作。Dynagen 是用于连接到 Dynamips 程序的调试界面，由于其界面与 DOS 界面类似，所以在 GNS3 中并不常用。在使用中我们经常会关闭 Console、Captures、Topology Summary 这些窗口，使得整个工作区界面更加整洁。

1.3.2 配置一台路由器

我们尝试着搭建最简单的拓扑，在节点类型中单击使用的 IOS 文件对应的路由器图标。这里我们以 C3640 为例。注意，只有当我们在 IOS 路径中加入其系统镜像，才能高亮这些图标，并且拖拽出来，这里我们只看到 C3600 图标高亮。从左侧面板拖动对应的路由器图标到中间的工作区，这样我们就有一个路由器可以配置了。右键单击路由器，在菜单中选择"配置"，如图 1-13 所示。

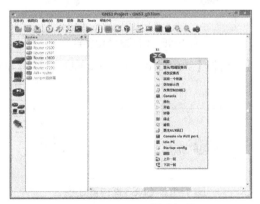

图 1-13　GNS3 搭建拓扑

在弹出的窗口中选择 R1，然后选择 Slots 标签，单击 Slot0 后面的下拉框，选择一个包括 FE 接口的板卡。这就会给路由器添加一个带快速以太网接口的板卡。接着

【拓扑】
Topology，网络拓扑是网络设备之间相互连接的结构示意图，可以包含路由器、交换机、主机、线缆等。

【接口模块】
网络设备的接口采用模块化设计，根据客户对接口的需求，增加不同的接口模块。C3640 这款路由器支持各种以太网和广域网接口模块。

在 Slot1 后面的下拉框中选择 NM-1FE-TX(1 个快速以太网口模块)、NM-1E（1 个以太网口模块）、NM-4E（4 个以太网口模块）、NM-16ESW（16 个快速以太网交换模块）、NM-4T（4 个广域网串口模块），这就会给路由器增加相应的端口。这里我们选择 NM-1FE-TX 模块，单击"OK"按钮。

右键单击路由器，选择"开始"。再次右键单击路由器，选择 Console，将会弹出一个 Console 窗口。或许我们需要在 Console 窗口中按回车键初始化路由器。过一小段时间，虚拟的路由器就启动完成了。也可以单击工具栏的 ▶ 按钮来启动全部机器，不过这样对电脑的性能要求比较高，建议一台一台地启动。在 GNS3 中路由器的命名以 R 开头。如果想修改路由器的名称，则可以在右键菜单中选择 "Change the hostname"。在下面的例子中使用默认名称，如图 1-14 所示。

图 1-14 GNS3 接口模块配置

1.3.3 连接两台路由器

①按照之前的方法，再连接一台路由器，配置一个一样的接口模块。之后单击 按钮连接两台设备，再单击 R1 路由器，会弹出我们之前配置的接口，选择 f0/0 接口，再单击 R2 的 f0/0，如图 1-15 所示。

【接口标识】
"F0/0" 这个接口标识由三部分组成，即接口类型、接口模块和接口序号。其中"F"代表 fastethernet，前面的"0"代表第一个模块，后面的"0"代表第一个模块的第一个接口。

图 1-15 GNS3 路由器连接

②单击【查看】，选择【show/hide interface lables】，打开显示接口名功能。如图 1-16 所示。

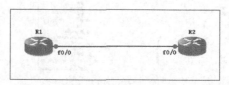

图 1-16 GNS3 路由器连接 2

1.3.4 通过 SecureCRT 管理设备

① SecureCRT 初始安装

SecureCRT 是网络工程师用的最多的一款终端登录软件，可以为我们提供命令行调试界面。GNS3 为我们提供了路由器和交换机等模拟平台，而 SecureCRT 则可以让我们登录这个平台，对 GNS3 上面的设备进行命令调试。SecureCRT 支持 Telnet、SSH 和 Serial 等协议。可以到其官网（http://www.vandyke.com/）下载最新版本，下载完成后安装并打开程序，如图 1-17 所示。

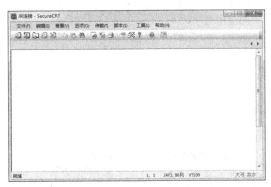

图 1-17　SecureCRT 启动界面

②通过 SecureCRT 登录路由器

默认所有的连接都以红色表示。开启路由器，启动后所有的连接都变为绿色，之后我们要通过 SecureCRT 来管理这两台设备。先打开 CRT 软件，第一次启动时会显示"快速连接"，如图 1-18 所示。SecureCRT 协议支持 SSHv1、SSHv2、Telnet、Serial 等，这里我们选 Telnet，主机名填 127.0.0.1（代表电脑本地），端口号为 2101（GNS3 默认拖出来的第一台路由器的端口为 2101，其余的依次递增），最后单击连接与连到路由器 R1、R2 的连接方法一样，如图 1-19 所示。

图 1-18　SecureCRT 快速连接

我们可以在 SecureCRT 界面上来配置路由器 R1 和 R2。
在 R1 上配置接口 IP 地址，如下所示。

```
R1#config terminal
R1(config)#interface f0/0
R1(config-if)#no shutdown
```

【终端登录】
终端登录即用户接入设备的动作，可以通过 Windows 系统自带的超级终端软件或者这里介绍的 SecureCRT 来实现。SecureCRT 是终端登录软件里面的佼佼者，能支持多协议多标签管理。

【Telnet】
远程登录协议，可实现设备远程管理，基于明文传输，应用非常广泛。

【SSH】
Telnet 协议的升级版，可以实现基于加密的设备远程管理。

【Serial】
串行协议，是电脑通过 Console 线近端管理网络设备的协议。

【127.0.0.1】
本地环回地址，代表电脑本地。也可以采用 127.0.0.2 或其他 127 开头的地址作为环回地址。

```
R1(config-if)#ip add 12.1.1.1 255.255.255.0
R1(config-if)#end
```

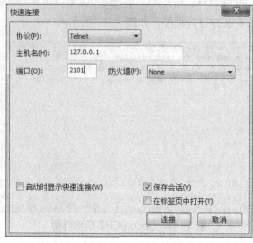

图 1-19　SecureCRT 快速连接 2

在 R2 上配置接口 IP 地址。

```
R2#config terminal
R2(config)#interface f0/0
R2(config-if)#no shutdown
R2(config-if)#ip add 12.1.1.2 255.255.255.0
R2(config-if)#end
```

1.3.5　使用 wireshark 抓取 ping 包

右键单击两个路由器之间的线，选择"开始抓包"，弹出一个对话框，如图1-20所示。

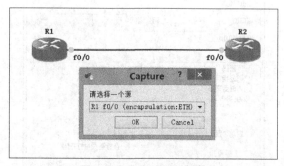

图 1-20　Wireshark 抓包

选择一个接口，这里可以选择 R1 的 f0/0 和 R2 的 f0/0，默认就可以了，单击"OK"按钮，会自动开启抓包软件 wireshark。注意，如果安装 wireshark 的时候改变了 wireshark 的安装目录，还需更改首选项中 Capture 的设置。

在 SecureCRT 中用 ping 命令 ping 对方路由器的 IP 地址，如下所示。

```
R1#ping 12.1.1.2
Type escape sequence to abort.
Sending 5, 100-byte ICMP Echos to 12.1.1.1, timeout is 2 seconds:
```

【抓包】

抓包即通过工具或软件抓取设备通信间的数据。网络行业里面一般采用 wireshark、sniffer、Omnipeek 等抓包软件。抓包可以帮助我们更好的理解数据底层通信机制。

【Ping】

Ping 是测试连通性的基本工具，采用 ICMP 协议进行工作。

```
.!!!!
Success rate is 80 percent (4/5), round-trip min/avg/max = 16/71/96 ms
```

回到 wireshark 查看刚刚 ping 的数据包,如图 1-21 所示。

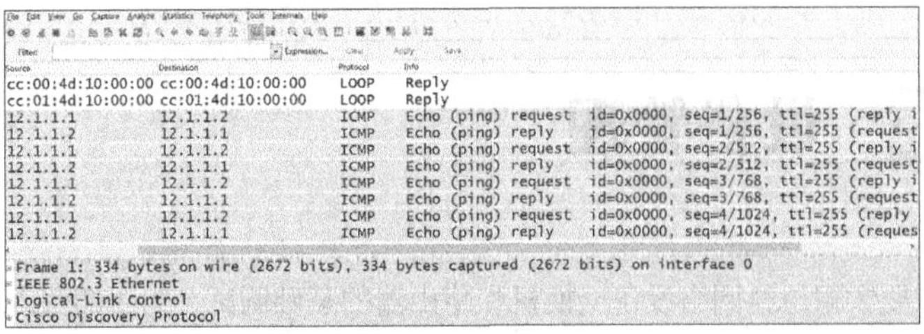

图 1-21　Wireshark 抓包 2

第 2 章　设备管理

本章主要学习思科网络设备交换机和路由器的基础调试，为更加深入地学习后续内容做铺垫。在本章中，我们将学习到基础的命令调试，了解常用的命令操作模式，掌握如何配置接口的 IP 地址，并且能对设备进行进阶的密码设置与恢复、操作系统备份与恢复等。以下为本章导航图。

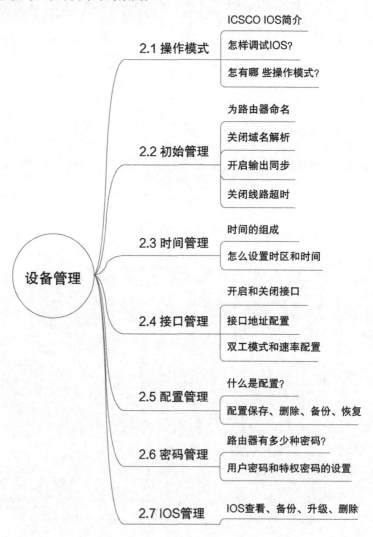

2.1 操作模式

实验目的:

1. 掌握路由器的几种操作模式。
2. 能在不同操作模式下进行切换。
3. 理解不同操作模式的功能和区别。

实验拓扑:

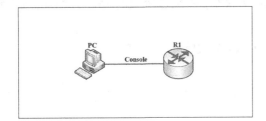

实验随手记:

_____

_____

_____

_____

_____

_____

实验原理:

1. Cisco IOS 简介

IOS 全称为 Internetwork Operating System,即网际操作系统,用于对思科公司的路由器和交换机进行操作,类似运行在个人电脑上的操作系统 Windows<例如 Windows 8>,其中 Win8 采用图形化界面对 PC 进行操作,而 IOS 则采用命令行界面对路由器和交换机进行操作。

2. IOS 操作模式

为了更好地对网络设备进行操作,需要熟悉和理解 IOS 的操作模式。IOS 的操作模式主要分为以下 3 种。

①用户模式 <user mode>:用于查看设备的基本信息。

②特权模式 <privilege mode>:用于查看设备更详细的信息,并可以实现基本的设备管理功能,例如配置文件和 IOS 镜像文件的管理。

③配置模式 <configuration mode>:此模式也称为全局模式,可以实现对设备的配置,如主机名修改、密码创建、IP 地址配置等。

3 种模式的设定是按照安全权限的高低等级来进行区分的,最低等级为用户模式,

【思科简介】
思科公司是目前全球互联网技术的领导者,成立于 1984 年 12 月,创始人是斯坦福大学的一对教师夫妇。全球第一台真正意义上的多协议路由器由其开发并使用于斯坦福大学校园网。全球互联网上有超过 80% 的信息流量要经过思科的网络设备如路由器和交换机传递。

【IOS 背景】
思科的网络操作系统起源于 UNIX 系统,最早由威廉·伊格在 1986 年编写。他是思科操作系统的真正发明者。

【Console 介绍】
Console 口是路由器、交换机、防火墙等。
网络设备的管理接口,与常用网线口一样采用 RJ45 接口。我们可以采用 Console 线缆,也叫控制线,用 PC 连接网络设备进行设备调试。

【日志信息】
当我们对设备进行命令调试时,经常会弹出一些系统信息,此为系统日志信息,即 syslog,有了系统日志信息,当设备的软件或者硬件发生变动的时候,syslog 都会告知我们,以方便我们排错。

【常用快捷键】
① <Ctrl+A> 将光标移动到命令最前。
② <Ctrl+E> 将光标移动到命令行

最高等级为配置模式,等级越高能拥有的命令越多,能实现的功能也越多。3 种模式刚好对应普通用户、管理员、超级管理员这 3 种角色。

实验步骤:

1. 通过 Console 连接 PC 和路由器,进入 SecureCRT 终端界面,如下所示。

```
Router>
// "Router" 是路由器的默认名字,可通过命令修改
// ">" 尖角号表示现在处于用户模式下
```

用户模式,是 3 种操作模式中权限最低的模式,所拥有的命令最少,只能做基本的设备管理。

2. 从用户模式切换到特权模式,如下所示。

```
Router>enable
// 输入完命令之后,要按回车,命令才会生效。命令 "enable" 的效果是进入特权模式
Router#
// "#" 井号表示现在处于特权模式下
```

在用户模式输入 "enable",即 "使能" 后进入特权模式,特权模式可以对设备进行高级管理,例如能查看设备大部分的运行。

3. 从特权模式切换到配置模式,如下所示。

```
Router#configure terminal
// 用这条命令进入 "配置模式"。一般为了操作方便,可缩写为 "conf t"
Enter configuration commands, one per line.  End with CNTL/Z.
// 以上为系统日志信息,表示的是:输入配置命令,一行一个命令,按 Ctrl+Z 结束
Router(config)#
// "(config)#" 表示现在处于 "配置模式" 下
```

如果说特权模式的权限相当于管理员,则配置模式便是超级管理员。配置模式能对设备进行部署,例如接口、线路、协议等配置。

4. 从配置模式切换到特权模式。

①通过快捷键 Ctrl+Z 实现,如下所示。

```
Router(config)#^Z
// 如上面提到的,按 Ctrl+Z 退出配置模式返回特权模式
Router#
// 现在处于特权模式下
```

②通过命令 exit 实现,如下所示。

```
Router(config)#exit
// 当然,输入 exit 也可以退出配置模式返回特权模式
Router#
```

③通过命令 end 实现,如下所示。

```
Router(config)#end
// 无论在哪一层，end 指令都可以直接退到特权模式下
Router#
```

其中 exit 和 end 的区别在于，前者只能一级一级跳到特权模式，后者则可以直接跳到特权模式，如下所示。

```
Router(config)#int f0/0
// 在配置模式下，通过此命令进入接口模式，IP 地址便需要在这种模式下配置
Router(config-if)#exit
// 退出端口配置模式，回到配置模式
Router(config)#
Router(config)#int f0/0
Router(config-if)#end
// 或者直接用 end 退回到特权模式
Router#
```

5. 从特权模式切换到用户模式。

①通过 exit 实现，如下所示。

```
Router#exit
//exit 是无论什么模式下都能使用的，而且无论在哪种模式都能回退一层
Router>
```

②通过 quit 实现，如下所示。

```
Router#quit
Router>
```

3 种设备操作模式组成 Cisco IOS 的安全管理框架，不同的模式对应不同的安全级别，从"普通用户"到"管理员"再到"超级管理员"，熟练掌握这 3 种操作模式对于后续的设备管理和网络部署非常重要。此实验完成。

最后。
③ <Ctrl+Z> 此命令用于从其他模式如接口模式或全局模式之间退到特权模式，作用跟 <end> 命令一样。
④ < ↑ ↓ > 上下键经常使用的，用于查看历史命令，减少重复输入。

2.2 初始管理

实验目的：
1. 掌握路由器的命名。
2. 掌握域名解析功能的关闭。
3. 掌握线路下输出同步开启和线路超时关闭。

实验拓扑：

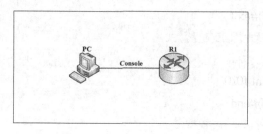

实验随手记：
_____
_____
_____
_____
_____

实验原理：

当我们第一次接触网络设备时，需要对设备进行初始化配置，例如给路由器配置主机名，方便后续整个实验的配置和管理。本实验在原有拓扑的基础上进行操作。

实验步骤：

1. 为路由器定义主机名，如下所示。

```
Router(config)#hostname R1
//hostname 后面必须有一个"空格"，之后加入我们为设备起的名字
R1(config)#
// 修改后，现在路由器名字成功显示为刚刚修改的那个 "R1" 了
```

为设备命名是初始化设备之后的首要工作，默认所有路由器的名字都是 Router，命名之后可以在网络中唯一区分这台设备，方便管理。

2. 关闭域名解析，如下所示。

```
R1(config)#no ip domain-lookup
// 配置模式下输入，关闭"域名解析"
```

【主机名】
为路由器设置主机名就如同为我们的电脑设置昵称。设备的主机名是工程环境下必须部署的，用于更加方便地管理网络。例如我们可以给校园网宿舍楼14栋接入层交换机命名为14AS1,14AS2,14AS3……代表宿舍楼14栋的几台接入层交换机。

默认情况下，当管理员敲错命令或者命令在本地无法查找时，路由器会触发域名解析功能，并且在域名解析过程中，管理员一般需要等待整个域名解析过程结束才能重新输入命令，如下所示。

```
R1#CCNALAB
Translating "CCNALAB"...domain server (255.255.255.255)
......
// 当键入的命令无法在本地查找时，则 IOS 会把该命令当成域名，然后再通过 DNS 查找该主机，这个过程需要等待一段时间
```

一般情况下，做实验的时候直接关闭此功能或者通过 Ctrl+Shift+6 命名来中断此进程。

3. Console 线路下关闭线路超时，如下所示。

```
R1(config)#line console 0
// 进入控制台模式
R1(config-line)#exec-timeout 0 0
// 如果是 exec-timeout 0 120 就是 2 分钟无操作断开
```

IOS 为了安全起见，当管理员离开终端一段时间后，IOS 会自动从其他模式跳转到用户模式。一般做实验的时候，将时间改为 0 分 0 秒，则表示关闭超时。

4. Console 线路下开启输出同步，如下所示。

```
R1(config)#line console 0
R1(config-line)#logging synchronous
// 开启输出同步，即在输入命令时设置不会被系统日志消息打断
```

当管理员在终端上进行操作时，一般会出现很多系统日志，这些系统日志会打乱当前管理员的配置，如下所示。

```
R1(config)#i
*Mar  1 00:22:29.903: %SYS-5-CONFIG_I: Configured from console by consolent f0/0
```

为了防止这种干扰，一般开启输出同步。

初始化管理是做后续其他实验的基础，无论后续是什么实验，基本上以上几条命令都属于"标配"。此实验完成。

【域名解析】
域名解析是操作系统最基本的功能之一，用于对我们常用访问域名如 www.baidu.com 解析成 IP 115.239.210.27 再进行访问。而在 IOS 环境下，我们的错误指令也会触发此机制。

2.3 时间管理

实验目的：
1. 掌握路由器的时间管理。
2. 理解时区、时间等概念。

实验拓扑：

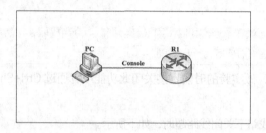

实验随手记：
_____
_____
_____
_____
_____

实验原理：

1. 时间的概念

思考一下，如果我们在路由器上做了一个网络策略，用于实现整个校园网络在凌晨 1 点到早上 5 点不能上互联网，那么怎么让路由器理解所谓的"时间"呢？这就需要我们给设备进行时间上的调试，所以网络设备上也有相应的"时钟"，用来与外部进行同步。网络设备的时间准确与否会很大程度上影响网络策略的执行效果，例如路由器的"时钟"与外部世界不同，并且策略已经在执行，就有可能造成白天校园网络无法上互联网，这种情况的出现会严重影响教务办公和学生的日常生活。

2. 时间的组成

时间由时区和具体时间共同组成，不同的国家或地区时区是不一样的，例如我国国内主要以北京时区东八区为主，IOS 操作系统默认时区为 UTC，UTC 全称为 Universal Time Coordinated，即世界统一时间，UTC+8= 北京时区。若没有修改时区，直接在设备上修改时间，则整体的时间与外部是不同的。所以在对设备进行时间管理时，需要同时修改时区和时间。

实验步骤：

1. 为路由器定义时区，如下所示。

【关于时间】
没有时区的时间是没有意义的，所以设置具体时间前要先设置时区。网络设备的时间可以通过手工调整，也可以通过 NTP 协议统一协调，一般工程环境中通过部署 NTP 服务器来统一时间。

```
R1(config)#clock timezone BJ +8
// 配置模式下输入，设置时区为东八区，BJ 是为这个时区起的名字
```

为设备设置具体时间之前要先设置时区，因为不同时区的时间是有差异的。中国的标准时区是东八区。

2. 为路由器定义时间，如下所示。

```
R1#clock set 10:30:00 1 JULY 2013
// 注意，这个是在特权模式下配置的，日期的格式是"日－月－年"
```

路由器的时间可以精确到秒。时区加上当地具体时间才是完整准确的时间。

3. 查看设备时间，如下所示。

```
R1#show clock
10:30:39.195 BJ Mon Jul 1 2013
```

时间对于网络管理来说是一个非常重要的属性。如果一个设备没有准确的时间，则管理员无法判断此设备的系统日志或其他报错信息在哪个具体时间点，这样对于网络故障排除非常不利。除了用手工方式为设备配置时间外，一般都采用 NTP 协议来进行全局设备时间管理。此实验完成。

2.4 接口管理

实验目的：
1. 掌握接口的开启和关闭。
2. 掌握接口地址和描述符配置。
3. 掌握接口双工模式和速率配置。

实验拓扑：

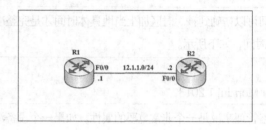

实验随手记：

_____
_____
_____
_____
_____
_____

实验原理：

1. 接口的概念

网络设备用于实现信息数据流的传递，整个互联网由无数的路由器、交换机、防火墙等设备组成，用于实现信息和数据的交换和传输。而接口是掌管数据出入的大门，数据出还是入，以及数据出入的速度等都要取决于接口的属性。这就需要我们对接口进行管理，如接口的关闭开启、接口的双工模式、接口的速率等。

2. 接口的属性

①双工模式：以太网和快速以太网接口的双工模式分为两种，一种为半双工 <half-duplex>；另一种为全双工 <full duplex>，不同双工模式对于网络的吞吐量有很大的影响。半双工模式下，同一时间内，接口要么收要么发，只能实现单方向的通信；而全双工模式下，同一时间内，接口既可以发送也可以接收。这样的话，若一秒钟内以接口传输数据为 100Mbit/s 计算，则半双工模式下，接收方和发送方的数据吞吐量为 100Mbit/s，同样的速率下，全双工的吞吐量可以达到 200Mbit/s。目前主流的双工模式是全双工模式。

②接口速率：以太网和快速以太网的接口速率分别为 10Mbit/s 和 100Mbit/s，一般接收方和发送方的接口速率需要一致。

【接口分类】
网络设备的接口总体可以分为两类，一类为管理接口，一类为通信接口。管理接口有 console 和 aux 两种口，一般常用 console 进行设备管理。通信接口主要分为局域网和广域网环境，例如以太网环境下有不同速度的以太网口，根据不同标准如 IEEE802.3/802.3u/802.3z 分别对应 10Mbit/s/100Mbit/s1000Mbit/s；广域网口实验环境下常用 Serial 串口。

2.4 接口管理

实验步骤：

1. 为路由器接口配置 IP 地址之前，先查看接口状态，如下所示。

```
R1#show ip interface brief
// 查看所有端口的 IP 状态信息，列出所有端口的简要情况
Interface      IP-Address OK? Method Status  Protocol
FastEthernet0/0unassigned YES unsetadministratively down  down
// administratively down 表示端口未开启，需执行 no shutdown 来打开端口
```

可以看到，默认情况下，路由器接口关闭，处于"down/down"状态，Status 标识接口物理状态，Protocol 标识接口链路或协议状态。一般当物理层处于 down 时，如网线没插好、网口被关闭等，链路或者协议状态也都是 down。

2. 为 R1 开启快速以太网口，并配置 IP 地址，如下所示。

```
R1(config)#int f0/0
// 进入接口模式
R1(config-if)#no shutdown
// 路由器的端口是关闭的，需要用"no shutdown"命令来打开
R1(config-if)#ip address 12.1.1.1 255.255.255.0
// 也可以简写成"ip add"后面空格接的是 IP 地址，之后再接子网掩码
R1(config-if)#exit
// 退出端口模式
```

查看接口信息，具体如下所示。

```
R1#show ip int brief
Interface  IP-Address OK? Metho  Status     Protocol
FastEthernet0/0 12.1.1.1   YES manual  up        down
```

可以看到，此时路由器接口处于"up/down"状态，当本地物理接口开启，但是链路对方有故障时或者协议故障时，则会出现此状况。

3. 为 R2 开启快速以太网口，并配置 IP 地址，如下所示。

```
R2(config)#int f0/0
R2(config-if)#no shutdown
R2(config-if)#ip address 12.1.1.2 255.255.255.0
R2(config-if)#exit
```

查看接口信息，具体如下所示。

```
R1#show ip int brief
Interface IP-Address OK?Metho Status Protocol
FastEthernet0/0 12.1.1.1 YES manual up  up


R2#show ip int brief
Interface    IP-Address OK? Method Status Protocol
```

【接口状态】
接口状态除了实验中出现的情况，还可能出现"up down"情况。这种情况下，一般是物理层开启，但是链路协议协商失败。

【接口地址】
同一链路的设备 IP 地址不能重叠，但是要保证在同一网段上，才能保证直连连通性。

FastEthernet0/0 12.1.1.2 YES manual up up

可以看到，此时双方接口状态处于"up/up"状态，表示接口正常。

4. 为接口配置描述符，如下所示。

R1(config)#int f0/0
R1(config-if)#description "Connect to CCNALAB*CCIE-Lab*R2"
//description 这个命令是接口标识命令，为接口加入描述信息
R1(config-if)#end

查看接口描述符信息，具体如下所示。

R1#show run int f0/0
// 查看具体接口的配置
Building configuration...
Current configuration : 143 bytes
interface FastEthernet0/0
 description "Connect to CCNALAB*CCIE-Lab*R2"

接口描述符可以帮助管理员更好地理解网络环境，方便故障排错。

5. 为接口配置双工模式，并设置接口速率，如下所示。

R1(config)#int f0/0
R1(config-if)#duplex full
// 设置接口为全双工模式
R1(config-if)#speed 100
// 设置接口速率为100Mbit/s
R1(config-if)#exit

查看接口双工模式，具体如下所示。

R1#show interfaces f0/0
// 显示该接口的具体物理信息
FastEthernet0/0 is up, line protocol is up
    Hardware is AmdFE, address is cc00.1edc.0000 (bia cc00.1edc.0000)
    Description: "Connect to CCNALAB*CCIE-Lab*R2"
    Internet address is 12.1.1.1/24
    MTU 1500 bytes, BW 100000 Kbit, DLY 100 usec,
        reliability 255/255, txload 1/255, rxload 1/255
    Encapsulation ARPA, loopback not set
    Keepalive set (10 sec)
    Full-duplex, 100Mb/s, 100BaseTX/FX
    ARP type: ARPA, ARP Timeout 04:00:00
    Last input 00:00:04, output 00:00:02, output hang never
    Last clearing of "show interface" counters never

2.4 接口管理

```
Input queue: 0/75/0/0 (size/max/drops/flushes); Total output drops: 0
Queueing strategy: fifo
Output queue: 0/40 (size/max)
5 minute input rate 0 bits/sec, 0 packets/sec
5 minute output rate 0 bits/sec, 0 packets/sec
    39 packets input, 11998 bytes
    Received 39 broadcasts, 0 runts, 0 giants, 0 throttles
0 input errors, 0 CRC, 0 frame, 0 overrun, 0 ignored
    0 watchdog
    0 input packets with dribble condition detected
    411 packets output, 42281 bytes, 0 underruns
    0 output errors, 0 collisions, 7 interface resets
    0 babbles, 0 late collision, 0 deferred
    0 lost carrier, 0 no carrier
    0 output buffer failures, 0 output buffers swapped out
```

链路的双工模式和接口速率必须一致，一般快速以太网默认为全双工，100Mbps 速率。

6. 测试直连连通性，如下所示。

```
R1#ping 12.1.1.2
// 利用 "ping" 命令可以检查网络是否连通，可以很好地帮助我们分析和判定网络故障。
Type escape sequence to abort.
Sending 5, 100-byte ICMP Echos to 12.1.1.2, timeout is 2 seconds:
.!!!!
// "." 表示未连通， "!" 表示连通。思考，为什么第一次 PING 会出现这种情况？
Success rate is 80 percent (4/5), round-trip min/avg/max = 28/47/56 ms
```

此时，R1 和 R2 之间可以正常通信。

接口管理对于后续的高级协议部署非常重要，一般在运行其他协议之前，都要保证直连连通。此实验完成。

【.!!!!】
一般第一次 Ping 的时候都是这种提示，这是由 ARP 造成的。第一次 Ping 需要寻找邻居的 MAC 地址，所以有一个 ARP 请求回应的延迟。之后再 Ping 则是 5 个感叹号。

2.5 配置管理

实验目的：
1. 掌握配置保存、删除。
2. 掌握配置备份、恢复。
3. 掌握配置查看。

实验拓扑：

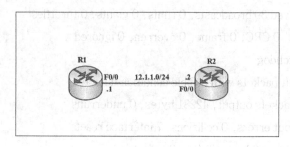

实验随手记：

_____
_____
_____
_____
_____

实验原理：

1. 配置文件的概念

任何网络设备例如交换机或者路由器都有配置文件，简称为配置，在路由器或交换机上一般称为 Configuration。配置文件一般存储在 NVRAM 或者 FLASH 下，里面包含着网络管理员或工程师对这些设备的调试指令集。在我们日常所使用的 Windows 操作系统中，我们对系统的修改 < 例如创建管理员、修改登录密码等 > 也会生成配置文件，这些配置文件一般被保存在系统盘固定的目录和文件下面。

2. 配置文件的管理

由于配置文件直接影响系统软件或硬件的运行状态，所以对配置文件的保存、删除、备份、恢复等就显得非常重要了。试想一下，校园网的核心交换机的硬件突然故障了，无法进行调试和修复，如果没有对核心交换机进行配置备份，那么这个故障就不仅仅是硬件故障了，而是在更换硬件设备后，还需要再花时间进行设备的指令调试。若在故障之前做好防范，对校园网络设备的所有配置进行备份，那么更换硬件设备后，将备份的配置导入新设备，即可正常运转。

【配置保存】
配置文件被加载到内存中运行，如果没有进行配置保存，则重启后机器还是处于初始化状态。没有配置的网络设备就相当于一块"板砖"！

2.5 配置管理

实验步骤:

1. 当设备配置完成后,需要保存配置在本地,防止重启后配置丢失。

①方法一,如下所示。

```
R1#copy running-config startup-config
// 将内存中的配置保存到 NVRAM 中,相当于我们电脑将配置储存在硬盘的特定文件中
Destination filename [startup-config]?
// 默认以 "startup-config" 为名字
Building configuration...
[OK]
```

②方法二,如下所示。

```
R1#write
// 保存配置的最常见写法
Building configuration...
[OK]
```

此时查看 NVRAM 中是否保存启动配置,如下所示。

```
R1#dir nvram:
// 显示磁盘目录命令

Directory of nvram:/

  124  -rw-      1007     <no date>  startup-config
  125  ----        24     <no date>  private-config
    1  -rw-         0     <no date>  ifIndex-table

129016 bytes total (126909 bytes free)
```

可以看到配置已经被保存。

2. 当配置出现错误时,需要删除配置,如下所示。

```
R1(config)#int f0/0
R1(config-if)#no ip address 12.1.1.1 255.255.255.0
R1(config-if)#exit
R1(config)#hostname R1
R1(config)#no hostname R1
// 用 "Ctrl+A" 则可以回到该行命令的最前端位置
Router(config)#
```

Cisco IOS 删除命令或配置直接在配置之前加上 "no" 即可,若不想一条一条地删除配置,则可以直接删除初始化配置并重启即可清空配置,如下所示。

【配置备份】
一般在网络管理员的电脑安装 FTP 和 TFTP 软件，然后将配置文件之间导入电脑，实现配置备份。

```
R1#write erase
// 直接擦除 NVRAM 的配置
Erasing the nvram filesystem will remove all configuration files! Continue? [confirm]
[OK]
Erase of nvram: complete
R1#reload
```

3. 为了防止设备故障而导致配置丢失，一般情况下，除了将配置保存在设备本地之外，还需要通过 FTP 或者 TFTP 将配置备份到服务器或者 PC 上。

① R2 开启 FTP 服务，如下所示。

```
R2(config)#ftp-server enable
// 开启 FTP 服务
R2(config)#ftp-server topdir flash:
// 定义 FTP 根目录为 FLASH:
R1#copy startup-config ftp:
// 将 R1 配置备份到 FTP 服务器
Address or name of remote host []? 12.1.1.2
// 输入目标的 IP 地址，即 FTP 服务器的地址
Destination filename [r1-confg]?
// 存储名为 "r1-confg"
Writing r1-confg !
509 bytes copied in 2.312 secs (220 bytes/sec)
```

② 在 R2 上查看 FLASH，如下所示。

```
R2#show flash:
// 查看 flash 下的目录文件
System flash directory:
File  Length   Name/status
 1    509      r1-confg
[576 bytes used, 8388028 available, 8388604 total]
8192K bytes of processor board System flash (Read/Write)
```

可以看到，此时 FTP 服务器上已经备份了 R1 的配置。

4. 当设备的配置丢失时，或者设备故障后换成新设备时，需要将配置重新加载进入，此时需要从其他已经备份好的 FTP 或者 TFTP 上恢复配置，如下所示。

```
R1#copy ftp: startup-config
// 从 ftp 上复制到本地 startup-config
Address or name of remote host []? 12.1.1.2
// 远端 FTP 服务器的地址
Source filename []? r1-confg
```

// 要复制的文件名字

Destination filename [startup-config]?

// 复制过来之后的名字

Accessing ftp://12.1.1.2/r1-confg...

Loading r1-confg

[OK - 509/4096 bytes]

[OK]

// 拷贝成功!

509 bytes copied in 9.672 secs (53 bytes/sec)

可以看到 R1 从 FTP 上加载了配置。

5. 查看设备配置。

① 查看系统初始化配置，如下所示。

R1#show startup-config

Using 509 out of 129016 bytes

!

version 12.4

service timestamps debug datetime msec

service timestamps log datetime msec

no service password-encryption

!

hostname R1

……

② 查看正在运行的配置，如下所示。

R1#show running-config

Building configuration...

Current configuration : 509 bytes

!

version 12.4

service timestamps debug datetime msec

service timestamps log datetime msec

no service password-encryption

!

hostname R1

……

设备的配置管理是管理员最常见的运维任务，保存和备份配置对于后续的网络运维可以起到很大的作用。此实验完成。

2.6 密码管理

实验目的：

1. 掌握 console 或 vty 下线路密码或用户名密码认证。
2. 掌握特权密码配置。
3. 掌握密码加密存储。

实验拓扑：

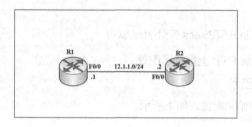

实验随手记：

_____
_____
_____
_____
_____

实验原理：

1. 密码的概念

我们经常会为个人的操作系统如 Windows 设置管理员登录密码，用于实现系统的安全性，防止被其他人随意登录并窃取重要资料等。同样的，我们需要为交换机和路由器设置登录密码,实现设备的安全管理。试想一下，一台在校园网络运行的路由器，例如是学校的互联网出口路由器，由于没有设置安全的登录权限，导致黑客随意登录，此时完全会有可能造成整个校园网络故障。比如，黑客对路由器的配置文件删除后重启，又或者将所有接口都关闭等。

2. 密码的设置

由于路由器和交换机有不同的操作模式，如用户模式、特权模式、配置模式，所以设置密码是针对不同的模式来进行的。例如要保护用户模式，则需要为 console（本地管理口）和 vty（远程登录口）设置密码；要保护特权模式，则需要设置特权密码，此处的密码相当于不同模式的安全锁，必须开了安全锁，管理员才有权限进入。

实验步骤：

1. 依据拓扑为设备配置 IP 地址,保证直连连通。在 R1 的 Console 下配置线路密码，

2.6 密码管理

如下所示。

```
R1(config)#line console 0
R1(config-line)#password CCNALAB
// 定义进入 Console 线路下的密码
R1(config-line)#login
// 调用密码，若没有此关键词，则密码无效
R1(config-line)#exit
```

此时，重新登录 R1，如下所示。

```
User Access Verification

Password:
// 输入密码时，不会有任何反应，因为是隐藏密码
R1>
```

同样方法在 VTY 线路下设置密码，如下所示。

```
R1(config)#line vty 0 15
//line vty 0 15 是进入 vty 端口，对 vty 端口进行配置
// VTY 是路由器远程登录的虚拟端口，015 表示可以同时打开 16 个会话
R1(config-line)#password CCNALAB
//Password 后空格，接上自己想设置的密码
R1(config-line)#login
// 必须要用此关键字来调用密码才起作用
R1(config-line)#exit
```

在 R2 上 Telnet R1，如下所示。

```
R2#telnet 12.1.1.1
//telnet 特定的 IP 地址，来远程控制该主机
Trying 12.1.1.1 ... Open


User Access Verification

Password:
R1>
```

通过以上实验，可以看到当我们在 Console 或者 VTY 线路下设置密码时，可以实现基本的安全管理。但是，此认证方式安全性比较薄弱，一般采用用户名密码的方式进行认证。

2.创建本地用户名数据库，并在 Console 和 VTY 下调用，如下所示。

【管理方式】
日常操作过程中，我们对网络设备的管理主要分为带内和带外管理两种方式，其中通过 console 线缆管理设备的方式称为带外管理，也称为近端管理；通过网线远程登录设备的管理方式称为带内管理，也称为远端管理。一般采用 Telnet 和 SSH 等协议进行远程登录。

```
R1(config)#username CCNALAB password cisco
// 设置用户名和对应密码，可以为不同的管理员设置不同的用户名和密码
R1(config)#line console 0
R1(config-line)#login local
// 调用本地的用户名和密码密码，倘若没有这个命令，则密码不生效
R1(config-line)#exit
R1(config)#line vty 0 15
R1(config-line)#login local
R1(config-line)#exit
```

此时，R1 重新通过 Console 登录，如下所示。

```
User Access Verification

Username: CCNALAB
// 输入用户名
Password:
// 输入密码
R1>
```

R2 上 Telnet R1，如下所示。

```
R2#telnet 12.1.1.1
Trying 12.1.1.1 ... Open
User Access Verification

Username: CCNALAB
Password:
R1>
```

通过用户名密码对 Console 和 VTY 线路下进行认证，可以实现分权管理，不同管理员的用户名可以不同，并且可以为不同的用户名定义不同的安全级别。

3. 为设备配置特权密码，如下所示。

```
R1(config)#enable password CCNALAB
// 特权密码意味着要输入密码后，才能进入到特权模式。
```

在 R1 上切换到用户模式并进入特权模式，如下所示。

```
R1>enable
Password:
// 同样在输入密码的时候，密码隐藏
R1#
```

此时进入特权模式需要特权密码，从而保证了特权模式的安全性。

4. 密码安全存储。默认情况下，密码明文存储，如下所示。

```
R1#show run
Building configuration...
……
enable passwordCCNALAB
username CCNALAB password 0 cisco
……
```

> 【密码安全】
> 为了实现更高强度的安全防护，需要设置安全度更高的密码，例如密码长度超过 8 位，并且采用密文存储。

为了实现更加安全的管理，可以通过两种方式实现密码加密存储。

① 将关键词 "password" 换成 "secret"，如下所示。

```
R1(config)#enable secret CCNALAB
R1(config)#username CCNALAB secret cisco
```

查看配置，如下所示。

```
R1#show run
Building configuration...
……
enable secret 5 $1$fruw$WuJpIdaCUBEFRy9mYD./Q/
username CCNALAB secret 5$1$bkZ1$0/uQVRwLexe/xK5QjebUb0
……
```

可以看到，此时密码被加密存储。

② 全局开启加密服务，如下所示。

```
R1(config)#service password-encryption
// 把所有的密码都加密，在 show run 的时候则密码都是被加密的
```

查看配置，如下所示。

```
R1#show run
Building configuration...
……
enable password 7 1139100B101B050B282B29
username CCNALAB password 7 14141B180F0B
……
```

同样可以看到，密码被加密存储。

密码管理是网络安全管理的基础，在现在网络安全越来越严峻的情况下，"锁好门"才是王道！此实验完成。

2.7 IOS 管理

实验目的：
掌握 IOS 查看、备份、升级、删除功能。

实验拓扑：

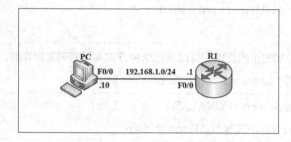

实验随手记：

_____
_____
_____
_____
_____

实验原理：
网络设备的操作系统 IOS 跟 Windows 类似，需要时常维护，例如系统的查看、备份、升级、删除等。操作系统是整个设备的灵魂，如果出现故障，便无法进行操作，设备就变成名副其实的"大砖头"，所以 IOS 的管理便显得更加重要了。

实验步骤：

1. 查看 R1 上 IOS 镜像，如下所示。

R1#show flash:

System flash directory:
File Length   Name/status
3   5571584   c2600-i-mz.122-28.bin

2. 备份 IOS。
在 PC 上架设 TFTP 服务器，并保证 PC 与 R1 直连连通，将 R1 上 IOS 镜像备份到 TFTP 上，如下所示。

【FLASH】
路由器或交换机的 FLASH，类似于电脑的硬盘，一般用于放置操作系统镜像，即 IOS。设备启动的时候，从 Flash 加载 IOS。在升级 IOS 的时候，要对原有 IOS 进行备份，并且确认 Flash 的大小是否支持新 IOS，若空间不够，则需要增加 Flash 闪存条。

```
R1#copy flash: tftp:
Source filename []? c2600-i-mz.122-28.bin
Address or name of remote host []? 192.168.1.10
Destination filename [c2600-i-mz.122-28.bin]?
Writing c2600-i-mz.122-28.bin
.!!!!!!!!!!!!!!!!!!!!!!!!!!!!!!!!!!!!!!!!!!!!!!!!!!!!!!!!!!!!!!!!!
[OK - 5571584 bytes]

5571584 bytes copied in 0.086 secs (64785000 bytes/sec)
```

当设备 IOS 损坏或者故障时，通过备份的 IOS 可以进行恢复。

3. 升级 IOS。

①若原有 IOS 不能支持网络需求，则需要对 IOS 进行升级，一般会将新的 IOS 放置在 TFTP Server 上，然后通过 TFTP 传递到设备上，如下所示。

```
R1#copy tftp: flash:
Address or name of remote host []? 192.168.1.10
Source filename []? c2600-advipservicesk9-mz.124-15.T1.bin
Destination filename [c2600-advipservicesk9-mz.124-15.T1.bin]?
Erase flash:before copying? [confirm]
Verifying checksum... OK (0xD589)
Accessing tftp://192.168.1.10/c2600-advipservicesk9-mz.124-15.T1.bin...
Loading…c2600-advipservicesk9-mz.124-15.T1.bin…from 192.168.1.10:
!!!!!!!!!!!!!!!!!!!
[OK - 33591768 bytes]
```

若想保留原有 IOS，则不要覆盖；若直接覆盖，则原有 IOS 被删除。

②查看 FLASH，如下所示。

```
R1#show flash:
System flash directory:
File  Length    Name/status
 4    33591768  c2600-advipservicesk9-mz.124-15.T1.bin
 3    5571584   c2600-i-mz.122-28.bin
```

③当有多个 IOS 时，则需要通过命令指定优先启动的 IOS，如下所示。

```
R1(config)#boot system flash:///c2600-advipservicesk9-mz.124-15.T1.bin
```

4. 删除 IOS 镜像。

若 FLASH 大小不够或者无需备份的 IOS，则可以删除 IOS，如下所示。

```
R1#delete flash:c2600-i-mz.122-28.bin
Delete filename [c2600-i-mz.122-28.bin]?
Delete flash:c2600-i-mz.122-28.bin? [confirm]
```

【TFTP 服务器架设】
安装光盘中的"tftp32"软件。

或者直接格式化 FLASH，如下所示。

```
R1#erase flash:
Erasing the flash filesystem will remove all files! Continue? [confirm]
Erasing device... eeeeeeeeeeeeeeeeeeeeeeeeeeeeeeeee ...erased
Erase of flash: complete
```

当然，格式化命令比较危险，慎用！

此实验完成。

第 3 章　TCP/IP 协议栈

本章主要学习 TCP/IP 协议栈，掌握常用协议如 ARP、IP、ICMP、UDP、TCP、DHCP、TELNET 的工作原理。除此之外，本章还通过抓包工具 Wireshark 对每个协议进行底层分组分析，使得我们可以更加清晰地了解协议的底层运行机制。以下为本章导航图：

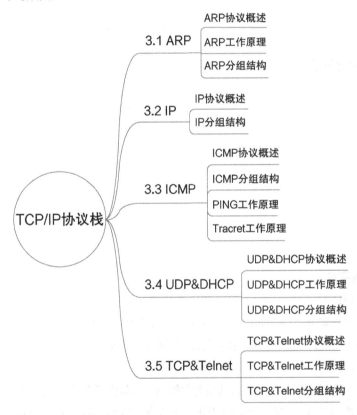

3.1 ARP

实验目的:

1. 通过 Wireshark 抓取 ARP 分组。
2. 掌握 ARP 的工作原理。

实验拓扑:

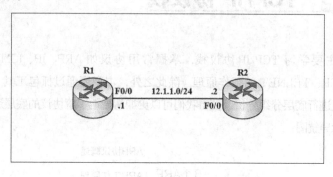

实验随手记:

_____

_____

_____

_____

_____

_____

实验原理:

1. ARP 简介

ARP（Address Resolution Protocol，地址解析协议），用于实现从 IP 地址到 MAC 地址的映射。它属于 OSI 七层模型中的网络层协议。

2. ARP 原理

在网络通信中，主机和主机通信的数据包需要依据 OSI 模型从上到下进行数据封装，当数据封装完成后，再向外发出。所以在局域网的通信中，不仅需要源目 IP 地址的封装，而且需要源目 MAC 的封装。一般情况下，上层应用程序更多关心 IP 地址而不关心 MAC 地址，所以需要通过 ARP 协议来获知目的主机的 MAC 地址，完成数据封装。其工作原理如图 3-1 所示。

【ARP】
ARP 由 IETF 互联网工程任务组于 1982 年 11 月在 RFC 826 中描述并发布。在 ARP 的基础上，衍生出了反向 ARP、代理 ARP、免费 ARP 等。

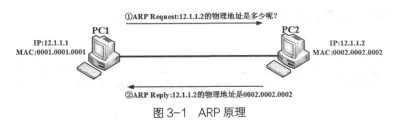

图 3-1　ARP 原理

从图中拓扑可以看出，PC1 需要与 PC2 进行通信，此时 PC1 向整个局域网发送 ARP Request，即 ARP 请求包，此请求包为二层广播包，目的地址为 ffff.ffff.ffff，保证同一局域网中的所有主机都能够收到此请求包。PC2 收到广播包请求后，向 PC1 发送单播的 ARP Reply 包，即 ARP 回复包。此后，PC1 将 PC2 的 IP 和 MAC 地址映射信息存储在本地 ARP 表项中，用于实现后续数据封装使用，如图 3-2 所示。

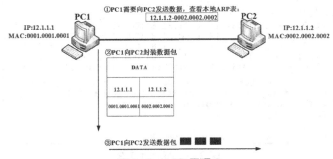

图 3-2　ARP 原理 2

从上图看出，当 PC1 需要访问 PC2 时，通过查看 ARP 映射表实现数据封装，之后再向外发送数据包。

实验步骤：

1. 通过 GNS3 搭建实验拓扑，采用两台路由器模拟主机，开启接口，并配置 IP 地址，如下所示。

```
R1(config)#int f0/0
R1(config-if)#no shutdown
R1(config-if)#ip address 12.1.1.1 255.255.255.0
R2(config)#int f0/0
R2(config-if)#no shutdown
R2(config-if)#ip add 12.1.1.2 255.255.255.0
```

2. 在 GNS3 上开启 wireshark 抓包，在 R1 和 R2 之间抓取 ARP 数据包。
①在 R1 和 R2 的链路上，单击右键，弹出"开始抓包"，如图 3-3 所示。

图 3-3　ARP 抓包

【数据封装】
数据包在网卡发送出去之前，需要在通信主机上经过应用层到物理层的七层封装和处理，最后才形成可以通信的包。

【广播】
网络通信中，有单播、组播和广播。单播为一对一通信，组播为一对多通信，而广播则是一对所有通信。

②弹出抓包对话框，单击"OK"按钮，如图 3-4 所示。

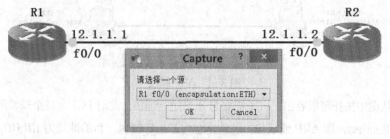

图 3-4　ARP 抓包 2

③此时 wireshark 对话框出现，并开始进行实时流量监控，如图 3-5 所示。

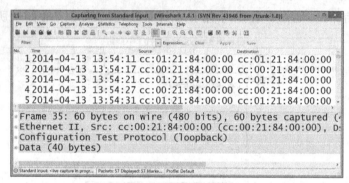

图 3-5　ARP 抓包 3

3. 在 R1 上 Ping R2 的地址，如下所示。

```
R1#ping 12.1.1.2
Type escape sequence to abort.
Sending 5, 100-byte ICMP Echos to 12.1.1.2, timeout is 2 seconds:
.!!!!
Success rate is 80 percent (4/5), round-trip min/avg/max = 32/38/44 ms
```

可以看到，此时 R1 Ping 通 R2。从实验结果来看，在发送 ICMP 数据包时（即 Ping 包），有一个包没有发送，所以显示为"."，而不是"！"。这说明在发送正常的 ICMP 数据包之前，有一个 ARP 的请求回复过程。

4. 在 wireshark 软件界面中观察数据包。

①通过拖拽 wireshark 右侧进度条，可以找到相对应的界面，如图 3-6 所示。

```
cc:00:21:84:00:00  Broadcast            ARP   Who has 12.1.1.2?  Tell 12.1.1.1
cc:01:21:84:00:00  cc:00:21:84:00:00    ARP   12.1.1.2 is at cc:01:21:84:00:00
cc:00:21:84:00:00  cc:00:21:84:00:00    LOOP  Reply
12.1.1.1           12.1.1.2             ICMP  Echo (ping) request  id=0x0000,
12.1.1.2           12.1.1.1             ICMP  Echo (ping) reply    id=0x0000,
12.1.1.1           12.1.1.2             ICMP  Echo (ping) request  id=0x0000,
12.1.1.2           12.1.1.1             ICMP  Echo (ping) reply    id=0x0000,
12.1.1.1           12.1.1.2             ICMP  Echo (ping) request  id=0x0000,
12.1.1.2           12.1.1.1             ICMP  Echo (ping) reply    id=0x0000,
12.1.1.1           12.1.1.2             ICMP  Echo (ping) request  id=0x0000,
12.1.1.2           12.1.1.1             ICMP  Echo (ping) reply    id=0x0000,
```

图 3-6　ARP 抓包 4

②接下来我们仔细分析 ARP 的分组内容，先将鼠标移至最上方的 ARP 请求包，得到图 3-7。

```
cc:00:21:84:00:00   Broadcast        ARP  Who has 12.1.1.2? Tell 12.1.1.1
cc:01:21:84:00:00   cc:00:21:84:00:00 ARP  12.1.1.2 is at cc:01:21:84:00:00
Frame 132: 60 bytes on wire (480 bits), 60 bytes captured (480 bits) on in
Ethernet II, Src: cc:00:21:84:00:00 (cc:00:21:84:00:00), Dst: Broadcast (f
Address Resolution Protocol (reply)
  Hardware type: Ethernet (1)
  Protocol type: IP (0x0800)
  Hardware size: 6
  Protocol size: 4
  Opcode: reply(2)
  Sender MAC address: cc:00:21:84:00:00 (cc:00:21:84:00:00)
  Sender IP address: 12.1.1.1 (12.1.1.1)
  Target MAC address: 00:00:00_00:00:00 (00:00:00:00:00:00)
  Target IP address: 12.1.1.2 (12.1.1.2)
```

图 3-7　ARP 抓包 5

以下表格详细解读了 ARP 请求包的内容。

字段	解释
Hardware type	硬件类型，标识链路层协议
Protocol type	协议类型，标识网络层协议
Hardware size	硬件地址大小，标识 MAC 地址长度
Protocol size	协议地址大小，标识 IP 地址长度
Opcode	操作代码，标识 ARP 数据包类型
Sender MAC address	发送者 MAC，此处为 R1 的 MAC
Sender IP address	发生者 IP，此处为 R1 的 IP
Target MAC address	目标 MAC，此处全 0 表示在请求
Target IP address	目标 IP，此处为 R2 的 IP

当发送 ARP 请求时，会将【Target MAC address】字段置空，代表本地没有，邻居收到之后，需要将此字段填充并返回。

③图 3-8 所示为 ARP 的回复包。

```
cc:00:21:84:00:00   Broadcast        ARP  Who has 12.1.1.2? Tell 12.1.1.1
cc:01:21:84:00:00   cc:00:21:84:00:00 ARP  12.1.1.2 is at cc:01:21:84:00:00
Frame 132: 60 bytes on wire (480 bits), 60 bytes captured (480 bits) on in
Ethernet II, Src: cc:00:21:84:00:00 (cc:00:21:84:00:00), Dst: Broadcast (f
Address Resolution Protocol (reply)
  Hardware type: Ethernet (1)
  Protocol type: IP (0x0800)
  Hardware size: 6
  Protocol size: 4
  Opcode: reply(2)
  Sender MAC address: cc:00:21:84:00:00 (cc:00:21:84:00:00)
  Sender IP address: 12.1.1.1 (12.1.1.1)
  Target MAC address: 00:00:00_00:00:00 (00:00:00:00:00:00)
  Target IP address: 12.1.1.2 (12.1.1.2)
```

图 3-8　ARP 抓包 6

此时可以看到，R2 将自己的 MAC 置于【Sender MAC address】字段中，完成整个 ARP 过程。

5. 在 R1 查看 ARP 表，如下所示。

```
R1#show arp
Protocol   Address     Age (min)   Hardware Addr    Type    Interface
Internet   12.1.1.1    –           cc00.2184.0000   ARPA    FastEthernet0/0
Internet   12.1.1.2    68          cc01.2184.0000   ARPA    FastEthernet0/0
```

可以看到，此时 R1 的 ARP 表除了有自己的地址映射信息，还具备 R2 的地址映射。此实验完成。

【ARP 表】
ARP 存储着 IP 与 MAC 地址的映射信息，用于后续的数据封装。在 Windows 系统下通过命令"arp -a"可以查看。

3.2 IP

实验目的：

1. 通过 Wireshark 抓取 IP 分组。
2. 通过 Wireshark 分析 IP 分组并理解其工作原理。

实验拓扑：

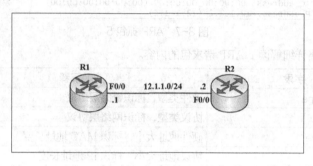

实验随手记：

_____

_____

_____

_____

_____

实验原理：

IP（Internet Protocol，互联网协议），是 TCP/IP 协议栈中一个非常重要的协议。IP 提供了面向无连接的不可靠传输的通信服务，除此之外，其提供的 IP 地址是目前互联网使用最广泛的寻址机制。

【IP】
IP 是在 RFC 文档 791 进行标准并发布的。IP 为网络中的通信设备提供地址信息，并尽力提供传输服务。

实验步骤：

1. 通过 GNS3 搭建实验拓扑，初始化两台设备，开启接口并配置 IP 地址，如下所示。

```
R1(config)#int f0/0
R1(config-if)#no shutdown
R1(config-if)#ip address 12.1.1.1 255.255.255.0

R2(config)#int f0/0
R2(config-if)#no shutdown
R2(config-if)#ip add 12.1.1.2 255.255.255.0
```

【面向无连接】
通信双方在传输数据之前没有建立任何连接会话。

2. 在 GNS3 上开启 wireshark 抓包，在 R1 和 R2 之间抓取 IP 数据包，并开始进行实时流量监控（步骤同上一个实验），此时 wireshark 对话框出现，如图 3-9 所示。

【不可靠传输】通信双方尽力传输数据，数据丢弃没法进行重传。

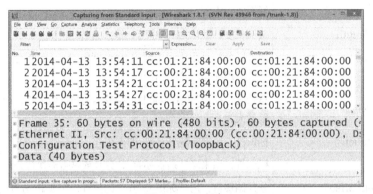

图 3-9　IP 抓包

3. 在 R1 上 Ping R2 的地址，如下所示。

R1#ping 12.1.1.2

Type escape sequence to abort.

Sending 5, 100-byte ICMP Echos to 12.1.1.2, timeout is 2 seconds:

.!!!!

Success rate is 80 percent (4/5), round-trip min/avg/max = 32/38/44 ms

4. 由于 ICMP 包基于 IP 协议，我们可以通过抓取 ICMP 分组来获得 IP 头部信息，如图 3-10 所示。

```
12.1.1.1        12.1.1.2        ICMP Echo (ping) request  id=0x0002, seq=0/
12.1.1.2        12.1.1.1        ICMP Echo (ping) reply    id=0x0002, seq=0/
12.1.1.1        12.1.1.2        ICMP Echo (ping) request  id=0x0002, seq=1/
12.1.1.2        12.1.1.1        ICMP Echo (ping) reply    id=0x0002, seq=1/
12.1.1.1        12.1.1.2        ICMP Echo (ping) request  id=0x0002, seq=2/
Frame 5: 114 bytes on wire (912 bits), 114 bytes captured (912 bits) on interfa
Ethernet II, Src: cc:00:21:84:00:00 (cc:00:21:84:00:00), Dst: cc:01:21:84:00:00
Internet Protocol Version 4, Src: 12.1.1.1 (12.1.1.1), Dst: 12.1.1.2 (12.1.1.2)
Internet Control Message Protocol
```

图 3-10　IP 抓包 2

单击 IP 头部，得到分组视图，如图 3-11 所示。

```
Frame 135: 114 bytes on wire (912 bits), 114 bytes captured (912 bits) on inter
Ethernet II, Src: cc:00:21:84:00:00 (cc:00:21:84:00:00), Dst: cc:01:21:84:00:00
Internet Protocol Version 4, Src: 12.1.1.1 (12.1.1.1), Dst: 12.1.1.2 (12.1.1.2)
  Version: 4
  Header length: 20 bytes
  Differentiated Services Field: 0x00 (DSCP 0x00: Default; ECN: 0x00: Not-ECT (N
  Total Length: 100
  Identification: 0x0001 (1)
  Flags: 0x00
  Fragment offset: 0
  Time to live: 255
  Protocol: ICMP (1)
  Header checksum: 0xa193 [correct]
  Source: 12.1.1.1 (12.1.1.1)
  Destination: 12.1.1.2 (12.1.1.2)
```

图 3-11　IP 抓包 3

以下表格详细解读了 IP 头部的内容。

【QoS】
Quality of Service，服务质量，用于实现流量控制技术。

【IP 分片】
当上层数据太大时，便需要对数据包进行分片。分片的时候会对不同的分片打上相同的标识符和偏移量，接收方根据标识符和偏移量对 IP 分片进行重组。

字段	解释
Version	版本号，标识 IP 的版本，目前 V4 版本地址已经枯竭，V6 慢慢成为主流
Header length	头部长度，默认为 20 字节，最大为 60 字节
Differentiated Services Field	服务区分符，用于为不同的 IP 数据包定义不同的服务质量，一般应用在 QoS 技术中
Total Length	总长度，标识 IP 头部加上上层数据的数据包大小，IP 包总长度最大为 65535 字节
Identification	标识符，用来实现 IP 分片的重组
Flags	标志符，用来确认是否还有 IP 分片或是否能执行分片
Fragment offset	分片偏移量，用于标识 IP 分片的位置，实现 IP 分片的重组
Time to live	生存时间，标识 IP 数据包还能生存多久。根据操作系统不同，TTL 默认值不同，每经过一个三层设备如路由器的处理，则 TTL 减去 1。当 TTL=0 时，则此数据包被丢弃
Protocol	协议号，标识 IP 上层应用。当上层协议为 ICMP 时，协议号为 1。TCP 协议号为 6，UDP 的协议号为 17
Header checksum	头部校验，用于检验 IP 数据包是否完整或被修改
Source	源 IP 地址，标识发送者 IP 地址，占用 32bit
Destination	目的 IP 地址，标识接收者 IP 地址，占用 32bit

5. 通过 wireshark 除了可以看到具体的分组内容，还可以看到分组所占用的长度。

①先将 wireshark 的分组长度选项【Packet Bytes】打开，如图 3-12 所示。

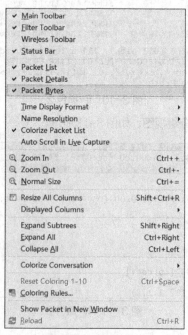

图 3-12　IP 抓包 4

此时可以看到 wireshark 的视图下面多了一个分组长度框，如图 3-13 所示。

```
12.1.1.1          12.1.1.2          ICMP Echo (ping) request   id=0x0002, seq=0/
12.1.1.2          12.1.1.1          ICMP Echo (ping) reply     id=0x0002, seq=0/
12.1.1.1          12.1.1.2          ICMP Echo (ping) request   id=0x0002, seq=1/
12.1.1.2          12.1.1.1          ICMP Echo (ping) reply     id=0x0002, seq=1/
12.1.1.1          12.1.1.2          ICMP Echo (ping) request   id=0x0002, seq=2/

Frame 5: 114 bytes on wire (912 bits), 114 bytes captured (912 bits) on interfa
Ethernet II, Src: cc:00:21:84:00:00 (cc:00:21:84:00:00), Dst: cc:01:21:84:00:00
Internet Protocol Version 4, Src: 12.1.1.1 (12.1.1.1), Dst: 12.1.1.2 (12.1.1.2)
Internet Control Message Protocol

008  00100001 10000010 00000000 00000000 00000100 00000000 01000101 00000000
010  00000000 01100100 00000000 00001010 00000000 00000000 11111111 00000001
018  10100001 10001010 00001100 00000001 00000001 00000001 00001100 00000001
020  00000001 00000010 00001000 00000000 11001011 10101101 00000000 00000010
028  00000000 00000000 00000000 00000000 00000000 00000000 00000000 10110110
030  10110001 11100100 10101011 11001101 10101011 11001101 10101011 11001101
038  10101011 11001101 10101011 11001101 10101011 11001101 10101011 11001101
040  10101011 11001101 10101011 11001101 10101011 11001101 10101011 11001101
```

图 3-13　IP 抓包 5

②单击不同的分组字段，便可以看到不同的分组长度，如图 3-14 所示。

```
Internet Protocol Version 4, Src: 12.1.1.1 (12.1.1.1), Dst: 12.1.1.2 (12.1.1.2)
    Version: 4
    Header length: 20 bytes
  Differentiated Services Field: 0x00 (DSCP 0x00: Default; ECN: 0x00: Not-ECT (N
    Total Length: 100
    Identification: 0x000a (10)
  Flags: 0x00
    Fragment offset: 0
    Time to live: 255
    Protocol: ICMP (1)
  Header checksum: 0xa18a [correct]
    Source: 12.1.1.1 (12.1.1.1)
    Destination: 12.1.1.2 (12.1.1.2)

0018   10100001 10001010 00001100 00000001 00000001 00000001 00001100 00000001
0020   00000001 00000010 00001000 00000000 11001011 10101101 00000000 00000010
0028   00000000 00000000 00000000 00000000 00000000 00000000 00000000 10110110
0030   10110001 11100100 10101011 11001101 10101011 11001101 10101011 11001101
```

图 3-14　IP 抓包 6

可以看到，IP 地址占用的长度为 32bit。此实验完成。

3.3 ICMP

实验目的：

1. 通过 Wireshark 抓取和分析 ICMP 分组。
2. 掌握 Ping 程序和 Tracert 程序的工作原理。

实验拓扑：

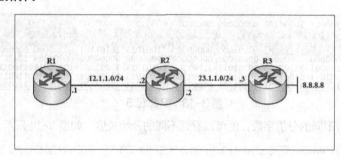

实验随手记：

_____
_____
_____
_____
_____

实验原理：

1. ICMP 简介

ICMP（Internet Control Message Protocol，互联网控制信息协议），用于实现 IP 网络层的连通性测试和差错报告。常用的 Ping 程序和 Tracert 程序便是基于 ICMP 协议进行开发的。

2. ICMP 分组

ICMP 根据类型值和代码值的组合，有非常多的分组类型，常用的 ICMP 分组有请求回显、回显应答、网络不可达、端口不可达、生存时间超时等。不同的 ICMP 分组应用在不同的环境，例如当采用 Ping 程序测试连通性时，则调用请求回显和回显应答分组；当采用 Tracert 程序时，则调用生存时间超时或端口不可达分组。

3. Ping 原理

Ping 程序是用于实现连通性测试的工具，图 3-15 解释了 Ping 的工作原理。

【ICMP】
ICMP 在 RFC 792 中标准化。在 IPv6 环境下，称为 ICMPv6。

【类型值&代码值】
类型值和代码值在 ICMP 中用于区分不同分组。ICMP 根据类型值和代码值，可以有几十种具体的 ICMP 分组。

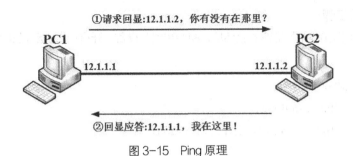

图 3-15　Ping 原理

从上图可以看到，PC1 在测试本地到 PC2 的连通性，所以向 PC2 发送 ICMP 的请求回显分组，PC2 收到请求之后，向 PC1 发送回显应答分组，此时 PC1 便可以得知 PC2 "在那里"。若 PC1 没有收到回显应答分组，则说明 PC1 与 PC2 链路有故障，如图 3-16 所示。

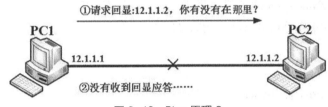

图 3-16　Ping 原理 2

4. Tracert 原理

Tracert 程序是用于实现链路追踪的工具，可以用来探测链路中的某个节点是否有问题，也可以用来分析网络拓扑结构。图 3-17 解释了 Tracert 的工作原理。

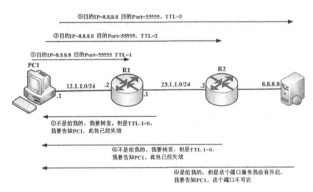

图 3-17　Tracert 原理

从上图可以看到，PC1 要追踪本地到目的服务器 8.8.8.8 的沿途路径，所以向外发送追踪包。追踪包是一个 UDP 分组，目的 IP 为 8.8.8.8，目的端口为 55555< 目的端口随机生成，但是数值非常高，大部分服务器没有提供此服务 >，TTL 从 1 开始不断递增。沿途的路由器收到追踪包之后，发现数据包不是给自己的，便开始查找路由表准备向外转发。当 TTL 减去 1 并且为 0 时，意味着此追踪包失效，路由器便需要返回一个 ICMP TTL 超时信息给源发送者；当数据包刚好到达目的服务器时，服务器不再将 TTL 减去 1，而是解封装到传输层，此时便看到 UDP 的高端口。如果本地没有这个端口服务时，则返回源通告者 ICMP 端口不可达信息，PC1 便可以根据返回的 IP 地址，收集沿途设备的 IP 列表，得知大概的网络拓扑结构。

第 3 章 TCP/IP 协议栈

实验步骤:

1. 根据实验拓扑搭建实验环境,初始化实验设备,开启接口并配置 IP 地址,如下所示。

R1(config)#int f0/0

R1(config-if)#no shutdown

R1(config-if)#ip address 12.1.1.1 255.255.255.0

R2(config)#int f0/0

R2(config-if)#no shutdown

R2(config-if)#ip add 12.1.1.2 255.255.255.0

R2(config)#int f1/0

R2(config-if)#no shutdown

R2(config-if)#ip address 23.1.1.2 255.255.255.0

R3(config)#int f0/0

R3(config-if)#no shutdown

R3(config-if)#ip address 23.1.1.3 255.255.255.0

R3(config)#int loopback 1

R3(config-if)#ip address 8.8.8.8 255.255.255.255

R3(config-if)#exit

2. 在 R1 和 R2 链路上开启 wireshark 抓包,并且在 R1 上 Ping R2,如图 3-18 所示。

R1#ping 12.1.1.2

Type escape sequence to abort.

Sending 5, 100-byte ICMP Echos to 12.1.1.2, timeout is 2 seconds:

.!!!!

Success rate is 80 percent (4/5), round-trip min/avg/max = 32/38/44 ms

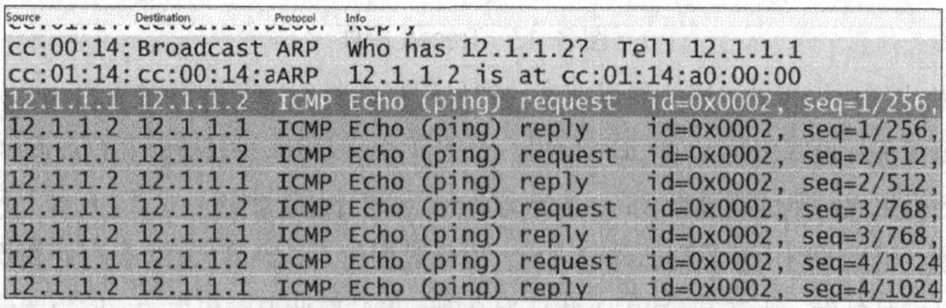

图 3-18 ICMP 抓包

3. 在 wireshark 软件界面中观察 ICMP 数据包,如图 3-19 所示。

```
Frame 87: 114 bytes on wire (912 bits), 114 bytes captured (912 bits) on interf
Ethernet II, Src: cc:00:14:a0:00:00 (cc:00:14:a0:00:00), Dst: cc:01:14:a0:00:00
Internet Protocol Version 4, Src: 12.1.1.1 (12.1.1.1), Dst: 12.1.1.2 (12.1.1.2)
Internet Control Message Protocol
  Type: 8 (Echo (ping) request)
  Code: 0
  Checksum: 0xa855 [correct]
  Identifier (BE): 2 (0x0002)
  Identifier (LE): 512 (0x0200)
  Sequence number (BE): 1 (0x0001)
  Sequence number (LE): 256 (0x0100)
  [Response In: 88]
  Data (72 bytes)
```

图 3-19　ICMP 抓包 2

从上图可以看到，R1 向 R2 发送 ICMP echo requtest，即回显请求分组，R2 向 R1 返回 ICMP echo reply，即回显应答分组，默认一次 Ping 有 5 个请求回复过程。

①单击此图中的 ICMP 请求回应分组，可以得到以下分组界面。

以下列表详细解读了 ICMP 请求回应分组的字段。

字段	解释
Type	类型值，标识 ICMP 分组类型
Code	代码值，标识 ICMP 分组类型的某一种具体分组
Checksum	校验和，用于检验数据包是否完整或是否被修改
Identifier	标识符，标识本进程。当同时与多个目的通信时，通过本字段来区分
Sequence Number	序列号，标识本地到目的的数据包序号，一般从序号 1 开始

②单击此图中的 ICMP 回显应答分组，可以得到以下分组界面，如图 3-20 所示。

```
Frame 88: 114 bytes on wire (912 bits), 114 bytes captured (912 bits) on interf
Ethernet II, Src: cc:01:14:a0:00:00 (cc:01:14:a0:00:00), Dst: cc:00:14:a0:00:00
Internet Protocol Version 4, Src: 12.1.1.2 (12.1.1.2), Dst: 12.1.1.1 (12.1.1.1)
Internet Control Message Protocol
  Type: 0 (Echo (ping) reply)
  Code: 0
  Checksum: 0xb055 [correct]
  Identifier (BE): 2 (0x0002)
  Identifier (LE): 512 (0x0200)
  Sequence number (BE): 1 (0x0001)
  Sequence number (LE): 256 (0x0100)
  [Response To: 87]
  [Response Time: 15.601 ms]
  Data (72 bytes)
```

图 3-20　ICMP 抓包 3

可以看到，回显应答与回显请求除了类型值不同外，其他字段基本一样。另外，两种分组在报文尾部夹带了时间戳信息，可以用于判断链路的来回延迟时间。

4. 在 R1、R2、R3 上部署 RIPv2 路由协议，保证全网连通性，如下所示。（路由协议的部署在后续路由章节有更详细的解读）

```
R1(config)#router rip
R1(config-router)#version 2
R1(config-router)#no auto-summary
R1(config-router)#network 12.0.0.0


R2(config)#router rip
R2(config-router)#version 2
R2(config-router)#no auto-summary
```

```
R2(config-router)#network 12.0.0.0
R2(config-router)#network 23.0.0.0


R3(config)#router rip
R3(config-router)#version 2


R3(config-router)#no auto-summary
R3(config-router)#network 23.0.0.0
R3(config-router)#network 8.0.0.0
```

5. 查看 R1 上的路由表,如下所示。

```
R1#show ip route
Codes: C – connected, S – static, R – RIP, M – mobile, B – BGP
       D – EIGRP, EX – EIGRP external, O – OSPF, IA – OSPF inter area
       N1 – OSPF NSSA external type 1, N2 – OSPF NSSA external type 2
       E1 – OSPF external type 1, E2 – OSPF external type 2
       i – IS–IS, su – IS–IS summary, L1 – IS–IS level-1, L2 – IS–IS level-2
       ia – IS–IS inter area, * – candidate default, U – per-user static route
       o – ODR, P – periodic downloaded static route

Gateway of last resort is not set

     23.0.0.0/24 is subnetted, 1 subnets
R       23.1.1.0 [120/1] via 12.1.1.2, 00:00:12, FastEthernet0/0
     8.0.0.0/32 is subnetted, 1 subnets
R       8.8.8.8 [120/2] via 12.1.1.2, 00:00:12, FastEthernet0/0
     12.0.0.0/24 is subnetted, 1 subnets
C       12.1.1.0 is directly connected, FastEthernet0/0
```

可以看到,R1 已经具有全网的路由。

6. 在 R1 和 R2 开启抓包,在 R1 上追踪 8.8.8.8,如下所示。

```
R1#traceroute 8.8.8.8


Type escape sequence to abort.
Tracing the route to 8.8.8.8


 1 12.1.1.2 40 msec 36 msec 24 msec
 2 23.1.1.3 36 msec 60 msec 40 msec
```

通过追踪 8.8.8.8,可以得到沿途路由器的 IP 地址。此时打开 wireshark 界面,如图 3-21 所示。

```
12.1.1.1  8.8.8.8    UDP  Source port: 49161  Destination port: tracerou
12.1.1.2  12.1.1.1   ICMP Time-to-live exceeded (Time to live exceeded
12.1.1.1  8.8.8.8    UDP  Source port: 49162  Destination port: 33435[M
12.1.1.2  12.1.1.1   ICMP Time-to-live exceeded (Time to live exceeded
12.1.1.1  8.8.8.8    UDP  Source port: 49163  Destination port: 33436[M
12.1.1.2  12.1.1.1   ICMP Time-to-live exceeded (Time to live exceeded
12.1.1.1  8.8.8.8    UDP  Source port: 49164  Destination port: 33437[M
23.1.1.3  12.1.1.1   ICMP Destination unreachable (Port unreachable)
12.1.1.1  8.8.8.8    UDP  Source port: 49165  Destination port: 33438[M
23.1.1.3  12.1.1.1   ICMP Destination unreachable (Port unreachable)
12.1.1.1  8.8.8.8    UDP  Source port: 49166  Destination port: 33439[M
23.1.1.3  12.1.1.1   ICMP Destination unreachable (Port unreachable)
```

图 3-21　ICMP 抓包 4

从上图可以看出，路由器每一跳发送 3 个 UDP 数据包，目的端口为高端口号，从 33435 开始；沿途设备 12.1.1.2 向源发生者 12.1.1.1 发送 TTL 超时信息，最终设备向源发生者返回 ICMP 端口不可达信息。

①打开 UDP 数据包，分组界面如图 3-22 所示。

```
Frame 3: 60 bytes on wire (480 bits), 60 bytes captured (480 bits) on interface
Ethernet II, Src: cc:00:21:78:00:00 (cc:00:21:78:00:00), Dst: cc:01:21:78:00:00
Internet Protocol Version 4, Src: 12.1.1.1 (12.1.1.1), Dst: 8.8.8.8 (8.8.8.8)
User Datagram Protocol, Src Port: 49161 (49161), Dst Port: traceroute (33434)
  Source port: 49161 (49161)
  Destination port: traceroute (33434)
  Length: 8
  Checksum: 0xa028 [validation disabled]
```

图 3-22　ICMP 抓包 5

②打开 ICMP TTL 超时包，分组界面如图 3-23 所示。

```
Frame 4: 70 bytes on wire (560 bits), 70 bytes captured (560 bits) on interface
Ethernet II, Src: cc:01:21:78:00:00 (cc:01:21:78:00:00), Dst: cc:00:21:78:00:00
Internet Protocol Version 4, Src: 12.1.1.2 (12.1.1.2), Dst: 12.1.1.1 (12.1.1.1)
Internet Control Message Protocol
  Type: 11 (Time-to-live exceeded)
  Code: 0 (Time to live exceeded in transit)
  Checksum: 0x122b [correct]
  Internet Protocol Version 4, Src: 12.1.1.1 (12.1.1.1), Dst: 8.8.8.8 (8.8.8.8)
  User Datagram Protocol, Src Port: 49161 (49161), Dst Port: traceroute (33434)
```

图 3-23　ICMP 抓包 6

③打开 ICMP 端口不可达包，分组界面如图 3-24 所示。

```
Frame 10: 70 bytes on wire (560 bits), 70 bytes captured (560 bits) on interfac
Ethernet II, Src: cc:01:21:78:00:00 (cc:01:21:78:00:00), Dst: cc:00:21:78:00:00
Internet Protocol Version 4, Src: 23.1.1.3 (23.1.1.3), Dst: 12.1.1.1 (12.1.1.1)
Internet Control Message Protocol
  Type: 3 (Destination unreachable)
  Code: 3 (Port unreachable)
  Checksum: 0x1a28 [correct]
  Internet Protocol Version 4, Src: 12.1.1.1 (12.1.1.1), Dst: 8.8.8.8 (8.8.8.8)
  User Datagram Protocol, Src Port: 49164 (49164), Dst Port: 33437 (33437)
```

图 3-24　ICMP 抓包 7

可以看到，追踪程序是根据 ICMP 返回的数据包源 IP 地址来获得网络的拓扑信息。此实验完成。

3.4 UDP&DHCP

实验目的：
1. 通过 Wireshark 抓取和分析 UDP 和 DHCP 分组。
2. 掌握 UDP 和 DHCP 的工作原理。

实验拓扑：

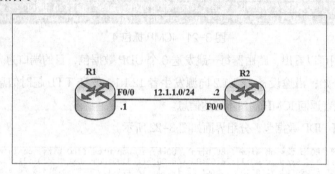

实验随手记：

_____
_____
_____
_____
_____
_____

实验原理：

1. UDP 简介

UDP（User Datagram Protocol，用户数据报协议），提供面向无连接的不可靠传输的通信服务，与 TCP 一起处于传输层。UDP 的特点是高效快速，常用的 UDP 应用有 DHCP、DNS、TFTP 等。

2. DHCP 简介

DHCP（Dynamic Host Configuration Protocol，动态主机配置协议），为网络中的设备提供动态 IP 地址信息，包括 IP 地址、网关、DNS 等。DHCP 可以使得整个网络的地址分配变得非常简单，大大减少了网络管理员的工作量。DHCP 基于 UDP，采用端口号为 67 和 68，其中 68 端口为 DHCP 客户端采用，67 端口为 DHCP 服务端采用。

3. DHCP 原理

DHCP 采用发现、提供、请求、确认 4 个分组完成整个 IP 地址的请求和分配过程，如图 3-25 所示。

【UDP】
UDP 在 IETF RFC 768 正式标准化。与 IP 一样，提供面向无连接不可靠的传输服务。

【DHCP】
DHCP 在 IETFRFC 2131 正式标准化。

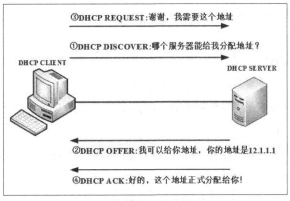

图 3-25　DHCP 原理

　　DHCP 客户端首先向局域网广播发送 DHCP 发现分组，源 IP 为 0.0.0.0，目的 IP 为 255.255.255.255，接收到发现分组的服务器会从地址池拿出一个 IP 出来，并返回 DHCP 提供分组给客户端，告知本服务器能够提供地址；当客户端收到提供分组之后，会正式向服务器发送请求分组，此时服务器发送确认分组正式将地址分配给客户端。整个 DHCP 地址分配过程采用广播包方式交互。若同时有几个 DHCP 服务器，则客户端能够收到多个提供分组，此时客户端优先选择最快到达本地的提供 <OFFER> 分组，并向其发起请求，在请求分组中夹带目的服务器的 IP 地址，其他服务器将地址收回。

【动态 IP】
通信设备所获取的地址具有租期，而不是永久的。租期一到便收回。

【网关】
网关一般指网络中的路由器，具备寻路功能，主机可以将数据丢给网关，网关根据路由表进行数据转发。

实验步骤：

1. 通过 GNS3 搭建实验拓扑，其中 R1 模拟 DHCP 客户端，R2 模拟 DHCP 服务端，并开启接口配置 IP 地址，如下所示。

```
R1(config)#int f0/0
R1(config-if)#no shutdown
R1(config-if)#ip address dhcp
// 在 R1 的接口开启 DHCP 地址自动获取
R2(config)#int f0/0
R2(config-if)#no shutdown
R2(config-if)#ip add 12.1.1.2 255.255.255.0
```

2. 在 R1 和 R2 上开启 wireshark 抓包，并且在 R2 上部署 DHCP 服务，如下所示。

```
R2(config)#ip dhcp pool DHCP
// 定义 DHCP 地址池，名字可以自定义
R2(dhcp-config)#network 12.1.1.0 /24
// 定义地址池的网段
R2(dhcp-config)#default-router 12.1.1.2
// 定义网关地址
```

```
R2(dhcp-config)#dns-server 8.8.8.8
```
// 定义 DNS 地址
```
R2(dhcp-config)#lease 7
```
// 定义地址租期，此处表示地址可用租期为 7 天

3. 在 R1 上查看是否获取到地址，如下所示。

```
R1#show ip int brief
Interface            IP-Address        OK? Method    Status        Protocol
FastEthernet0/0      12.1.1.1          YES DHCP     up            up
```

可以看到，此处 R1 已经通过 DHCP 获取到地址。

4. 打开 wireshark，观察 UDP 和 DHCP 分组，如图 3-26 所示。

```
Source          Destination       Protocol  Info
0.0.0.0         255.255.255.255   DHCP      DHCP Discover  - Transaction ID 0x2211
12.1.1.2        255.255.255.255   DHCP      DHCP Offer     - Transaction ID 0x2211
0.0.0.0         255.255.255.255   DHCP      DHCP Request   - Transaction ID 0x2211
12.1.1.2        255.255.255.255   DHCP      DHCP ACK       - Transaction ID 0x2211
```

图 3-26　DHCP 抓包

① 打开 DHCP 分组，打开 UDP 头部，如图 3-27 所示。

```
Frame 3: 618 bytes on wire (4944 bits), 618 bytes captured (4944 bits) on
Ethernet II, Src: cc:00:21:78:00:00 (cc:00:21:78:00:00), Dst: Broadcast (f
Internet Protocol Version 4, Src: 0.0.0.0 (0.0.0.0), Dst: 255.255.255.255
User Datagram Protocol, Src Port: bootpc (68), Dst Port: bootps (67)
  Source port: bootpc (68)
  Destination port: bootps (67)
  Length: 584
  Checksum: 0x0d03 [validation disabled]
```

图 3-27　DHCP 抓包 2

以下列表对 UDP 头部字段进行解读。

字段	解释
Source port	源端口，DHCP 客户端端口为 68
Destination port	目的端口，DHCP 服务端端口为 67
Length	长度，标识 UDP 头部和上层数据的总长度
Checksum	校验和，标识此数据包是否完整或被修改

可以看到，UDP 的头部非常简单，没有 TCP 的可靠传输或流量控制功能等。

② 打开 DHCP DISCOVER<发现>分组，如图 3-28 所示。

```
Frame 3: 618 bytes on wire (4944 bits), 618 bytes captured (4944 bits) on interface 0
Ethernet II, Src: cc:00:21:78:00:00 (cc:00:21:78:00:00), Dst: Broadcast (ff:ff:ff:ff:ff:ff)
Internet Protocol Version 4, Src: 0.0.0.0 (0.0.0.0), Dst: 255.255.255.255 (255.255.255.255)
User Datagram Protocol, Src Port: bootpc (68), Dst Port: bootps (67)
Bootstrap Protocol
  Message type: Boot Request (1)
  Hardware type: Ethernet
  Hardware address length: 6
  Hops: 0
  Transaction ID: 0x00002211
  Seconds elapsed: 0
  Bootp flags: 0x8000 (Broadcast)
  Client IP address: 0.0.0.0 (0.0.0.0)
  Your (client) IP address: 0.0.0.0 (0.0.0.0)
  Next server IP address: 0.0.0.0 (0.0.0.0)
  Relay agent IP address: 0.0.0.0 (0.0.0.0)
  Client MAC address: cc:00:21:78:00:00 (cc:00:21:78:00:00)
  Client hardware address padding: 00000000000000000000
  Server host name not given
  Boot file name not given
  Magic cookie: DHCP
  Option: (53) DHCP Message Type
  Option: (57) Maximum DHCP Message Size
  Option: (61) Client identifier
  Option: (12) Host Name
  Option: (55) Parameter Request List
  Option: (255) End
  Padding
```

图 3-28　DHCP 发现分组

以下列表详细解读了 DHCP 发现分组的重要字段。

字段	解释
Your (client) IP address	客户 IP 地址，客户端请求时为全 0，服务端返回时将地址填充进来
Client MAC address	客户 MAC 地址，客户端发起请求时必须填充此字段，服务端分配地址后将 MAC 与 IP 绑定在本地地址池中
Option 53	选项 53，标识 DHCP 分组类型
Option 57	选项 57，标识最大的 DHCP 分组长度
Option 61	选项 61，客户标识符
Option 12	选项 12，标识主机名
Option 55	选项 55，标识客户请求参数

③打开 DHCP OFFER（提供）分组，如图 3-29 所示。

```
Internet Protocol Version 4, Src: 12.1.1.2 (12.1.1.2), Dst: 255.255.255.255 (255.255.255.255)
User Datagram Protocol, Src Port: bootps (67), Dst Port: bootpc (68)
Bootstrap Protocol
  Message type: Boot Reply (2)
  Hardware type: Ethernet
  Hardware address length: 6
  Hops: 0
  Transaction ID: 0x0000165a
  Seconds elapsed: 0
  Bootp flags: 0x8000 (Broadcast)
  Client IP address: 0.0.0.0 (0.0.0.0)
  Your (client) IP address: 12.1.1.3 (12.1.1.3)
  Next server IP address: 0.0.0.0 (0.0.0.0)
  Relay agent IP address: 0.0.0.0 (0.0.0.0)
  Client MAC address: cc:00:2b:08:00:00 (cc:00:2b:08:00:00)
  Client hardware address padding: 00000000000000000000
  Server host name not given
  Boot file name not given
  Magic cookie: DHCP
  Option: (53) DHCP Message Type
  Option: (54) DHCP Server Identifier
  Option: (51) IP Address Lease Time
  Option: (58) Renewal Time Value
  Option: (59) Rebinding Time Value
  Option: (1) Subnet Mask
  Option: (3) Router
  Option: (6) Domain Name Server
  Option: (255) End
```

图 3-29　DHCP 提供分组

以下列表详细解读了 DHCP 提供分组的重要字段。

字段	解释
Option 54	选项 54，标识 DHCP 服务器的 IP 地址
Option 51	选项 51，标识 DHCP 地址租期
Option 1	选项 1，标识 IP 地址掩码
Option 3	选项 3，标识网关 IP 地址
Option 6	选项 6，标识 DNS 服务器地址

④打开 DHCP REQUEST（请求）分组，如图 3-30 所示。

```
Internet Protocol Version 4, Src: 0.0.0.0 (0.0.0.0), Dst: 255.255.255.255 (255.255.255.255)
User Datagram Protocol, Src Port: bootpc (68), Dst Port: bootps (67)
Bootstrap Protocol
  Message type: Boot Request (1)
  Hardware type: Ethernet
  Hardware address length: 6
  Hops: 0
  Transaction ID: 0x0000165a
  Seconds elapsed: 0
⊞ Bootp flags: 0x8000 (Broadcast)
  Client IP address: 0.0.0.0 (0.0.0.0)
  Your (client) IP address: 0.0.0.0 (0.0.0.0)
  Next server IP address: 0.0.0.0 (0.0.0.0)
  Relay agent IP address: 0.0.0.0 (0.0.0.0)
  Client MAC address: cc:00:2b:08:00:00 (cc:00:2b:08:00:00)
  Client hardware address padding: 00000000000000000000
  Server host name not given
  Boot file name not given
  Magic cookie: DHCP
⊞ Option: (53) DHCP Message Type
⊞ Option: (57) Maximum DHCP Message Size
⊞ Option: (61) Client identifier
⊞ Option: (54) DHCP Server Identifier
⊞ Option: (50) Requested IP Address
⊞ Option: (51) IP Address Lease Time
⊞ Option: (12) Host Name
⊞ Option: (55) Parameter Request List
```

图 3-30 DHCP 请求分组

以下表格详细解读了 DHCP 请求分组的重要字段。

字段	解释
Option 50	选项 50，标识客户端所请求的 IP 地址

⑤打开 DHCP ACK（应答）分组，如图 3-31 所示。

```
Internet Protocol Version 4, Src: 12.1.1.2 (12.1.1.2), Dst: 255.255.255.255 (255.255.255.255)
User Datagram Protocol, Src Port: bootps (67), Dst Port: bootpc (68)
Bootstrap Protocol
  Message type: Boot Reply (2)
  Hardware type: Ethernet
  Hardware address length: 6
  Hops: 0
  Transaction ID: 0x0000165a
  Seconds elapsed: 0
⊞ Bootp flags: 0x8000 (Broadcast)
  Client IP address: 0.0.0.0 (0.0.0.0)
  Your (client) IP address: 12.1.1.3 (12.1.1.3)
  Next server IP address: 0.0.0.0 (0.0.0.0)
  Relay agent IP address: 0.0.0.0 (0.0.0.0)
  Client MAC address: cc:00:2b:08:00:00 (cc:00:2b:08:00:00)
  Client hardware address padding: 00000000000000000000
  Server host name not given
  Boot file name not given
  Magic cookie: DHCP
⊞ Option: (53) DHCP Message Type
⊞ Option: (54) DHCP Server Identifier
⊞ Option: (51) IP Address Lease Time
⊞ Option: (58) Renewal Time Value
⊞ Option: (59) Rebinding Time Value
⊞ Option: (12) Host Name
⊞ Option: (1) Subnet Mask
⊞ Option: (3) Router
⊞ Option: (6) Domain Name Server
```

图 3-31 DHCP 应答分组

应答分组的选项值在之前有解读，此处略过。可以看到，DHCP 分组的大部分字段内容由选型值组成。此实验完成。

3.5 TCP&Telnet

实验目的：
1. 通过 Wireshark 抓取和分析 TCP 和 Telnet 分组。
2. 掌握 TCP 和 Telnet 的工作原理。

实验拓扑：

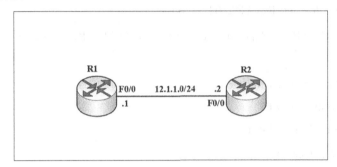

实验随手记：

_____
_____
_____
_____
_____

实验原理：

1. TCP 简介

TCP（Transmission Control Protocol，传输控制协议），提供面向连接的可靠传输服务，属于 OSI 七层模型的传输层协议，是 TCP/IP 协议栈中非常重要的协议。常用的基于 TCP 的协议有 HTTP、FTP、TELNET 等。

2. Telnet 简介

Telnet 即远程登录协议，可以为网络管理员提供非常方便的网管服务，只需要在服务器或网络设备上开启 Telnet 服务，便可以在本地远程访问服务器或网络设备，实现远程网管。Telnet 基于 TCP，采用端口号 23，并采用明文传输。本实验验证 TCP 的三次握手、可靠传输和四次挥手功能。

实验步骤：

1. 通过 GNS3 搭建实验拓扑，采用 R1 模拟 TELNET 客户端，R2 模拟 TELNET 服务端，将设备初始化，并开启接口配置 IP 地址，如下所示。

【TCP】
TCP 在 IETF RFC 793 中标准化。

【Telnet】
Telnet 在 IETF RFC 854 中标准化。

【面向连接】
通信双方在通信之前建立会话。

【可靠传输】
数据传输过程有确认信息，数据丢弃可以进行重传。

```
R1(config)#int f0/0
R1(config-if)#no shutdown
R1(config-if)#ip address 12.1.1.1 255.255.255.0

R2(config)#int f0/0
R2(config-if)#no shutdown
R2(config-if)#ip add 12.1.1.2 255.255.255.0
```

2. 在 R1 和 R2 上开启 wireshark 抓包，并且在 R2 上部署 TELNET 服务，如下所示。

```
R2(config)#username ccna secret cisco
R2(config)#line vty 0 15
R2(config-line)#login local
```

3. 在 R1 上远程登录 R2，如下所示。

```
R1#telnet 12.1.1.2
Trying 12.1.1.2 ... Open
User Access Verification


Username: ccna
Password:
R2>
```

4. 此时 R1 登录 R2，在 R2 上输入一些命令，例如查看版本号并退出（此处是为了验证 TELNET 的明文传输和 TCP 的四次关闭），如下所示。

```
R2>show version
Cisco IOS Software, 3600 Software (C3640-JK9O3S-M), Version 12.4(7a),
RELEASE SOFTWARE (fc3)
Technical Support: http://www.cisco.com/techsupport
Copyright (c) 1986-2006 by Cisco Systems, Inc.
Compiled Mon 24-Apr-06 23:37 by ssearch

……中间部分省略……
Configuration register is 0x2102
R2> exit
[Connection to 12.1.1.2 closed by foreign host]
R1#
```

5. 打开 Wireshark，查看 TCP 和 TELNET 分组，如图 3-32 所示。

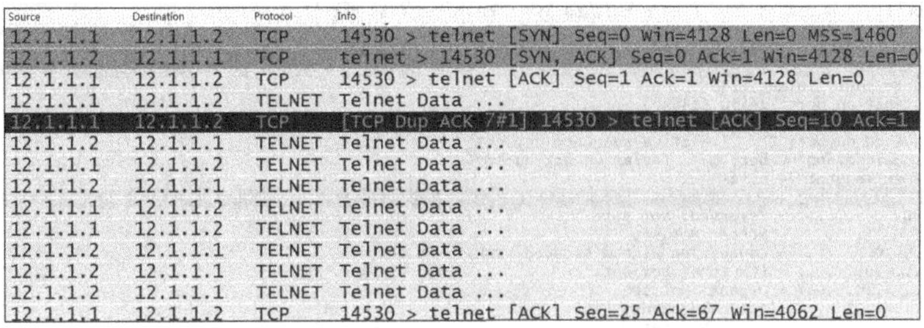

```
Source          Destination     Protocol    Info
12.1.1.1        12.1.1.2        TCP         14530 > telnet [SYN] Seq=0 Win=4128 Len=0 MSS=1460
12.1.1.2        12.1.1.1        TCP         telnet > 14530 [SYN, ACK] Seq=0 Ack=1 Win=4128 Len=0
12.1.1.1        12.1.1.2        TCP         14530 > telnet [ACK] Seq=1 Ack=1 Win=4128 Len=0
12.1.1.1        12.1.1.2        TELNET      Telnet Data ...
12.1.1.1        12.1.1.2        TCP         [TCP Dup ACK 7#1] 14530 > telnet [ACK] Seq=10 Ack=1
12.1.1.2        12.1.1.1        TELNET      Telnet Data ...
12.1.1.2        12.1.1.1        TELNET      Telnet Data ...
12.1.1.1        12.1.1.2        TELNET      Telnet Data ...
12.1.1.1        12.1.1.2        TELNET      Telnet Data ...
12.1.1.2        12.1.1.1        TELNET      Telnet Data ...
12.1.1.2        12.1.1.1        TELNET      Telnet Data ...
12.1.1.1        12.1.1.2        TCP         14530 > telnet [ACK] Seq=25 Ack=67 Win=4062 Len=0
```

图 3-32　TCP 和 TELNET 分组

① 分析 TCP 报文头部和三次握手原理，打开三次握手的第一个分组，如图 3-33 所示。

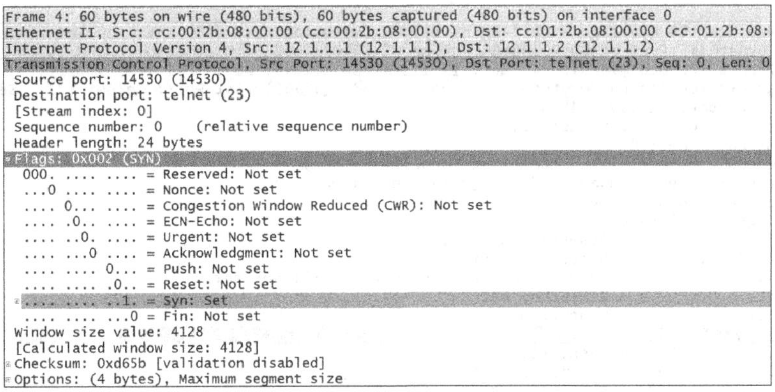

```
Frame 4: 60 bytes on wire (480 bits), 60 bytes captured (480 bits) on interface 0
Ethernet II, Src: cc:00:2b:08:00:00 (cc:00:2b:08:00:00), Dst: cc:01:2b:08:00:00 (cc:01:2b:08:
Internet Protocol Version 4, Src: 12.1.1.1 (12.1.1.1), Dst: 12.1.1.2 (12.1.1.2)
Transmission Control Protocol, Src Port: 14530 (14530), Dst Port: telnet (23), Seq: 0, Len: 0
  Source port: 14530 (14530)
  Destination port: telnet (23)
  [Stream index: 0]
  Sequence number: 0    (relative sequence number)
  Header length: 24 bytes
▼ Flags: 0x002 (SYN)
    000. .... .... = Reserved: Not set
    ...0 .... .... = Nonce: Not set
    .... 0... .... = Congestion Window Reduced (CWR): Not set
    .... .0.. .... = ECN-Echo: Not set
    .... ..0. .... = Urgent: Not set
    .... ...0 .... = Acknowledgment: Not set
    .... .... 0... = Push: Not set
    .... .... .0.. = Reset: Not set
    .... .... ..1. = Syn: Set
    .... .... ...0 = Fin: Not set
  Window size value: 4128
  [Calculated window size: 4128]
  Checksum: 0xd65b [validation disabled]
  Options: (4 bytes), Maximum segment size
```

图 3-33　三次握手的第一个 TCP 分组

以下列表对 TCP 头部的字段进行详细解读。

字段	解释
Source port	源端口，客户端端口随机生成
Destination port	目的端口，TELNET 服务端端口为 23
Sequence number	序列号，标识本地发送给通信接收方的第几个分组
Acknowledgment number	确认号，标识本地接收到第几个分组，在可靠情况下，确认号等于序列号加上数据包长度
Header length	头部长度，标识 TCP 头部长度
Flags	标志位，标识数据包的状态
Window size value	窗口大小，用于实现滑动窗口
Checksum	校验和，标识此数据包是否完整或被修改

SYN 位（同步位）打开，即表示客户端发起服务请求。打开三次握手的第二个 TCP 分组，如图 3-34 所示。

可以看到，第二个分组中，确认位从 0 变成 1，表示服务端接收服务请求。于此同时，由于通信是双向的，服务端也对客户端发起 SYN 同步请求，让其开启源端口，用于双向通信。打开三次握手的第三个 TCP 分组，如图 3-35 所示。

【滑动窗口】
通信双方根据滑动窗口控制接收的数据量。发送方发送的数据量不能超过接收方的窗口大小。

第 3 章 TCP/IP 协议栈

```
Frame 5: 60 bytes on wire (480 bits), 60 bytes captured (480 bits) on interface 0
Ethernet II, Src: cc:01:2b:08:00:00 (cc:01:2b:08:00:00), Dst: cc:00:2b:08:00:00 (cc:00:2b:08:00:00)
Internet Protocol Version 4, Src: 12.1.1.2 (12.1.1.2), Dst: 12.1.1.1 (12.1.1.1)
Transmission Control Protocol, Src Port: telnet (23), Dst Port: 14530 (14530), Seq: 0, Ack: 1, Len: 0
    Source port: telnet (23)
    Destination port: 14530 (14530)
    [Stream index: 0]
    Sequence number: 0    (relative sequence number)
    Acknowledgment number: 1    (relative ack number)
    Header length: 24 bytes
    Flags: 0x012 (SYN, ACK)
        000. .... .... = Reserved: Not set
        ...0 .... .... = Nonce: Not set
        .... 0... .... = Congestion Window Reduced (CWR): Not set
        .... .0.. .... = ECN-Echo: Not set
        .... ..0. .... = Urgent: Not set
        .... ...1 .... = Acknowledgment: Set
        .... .... 0... = Push: Not set
        .... .... .0.. = Reset: Not set
        .... .... ..1. = Syn: Set
        .... .... ...0 = Fin: Not set
    Window size value: 4128
    [Calculated window size: 4128]
    Checksum: 0x207d [validation disabled]
```

图 3-34　三次握手的第二个 TCP 分组

```
Frame 121: 60 bytes on wire (480 bits), 60 bytes captured (480 bits) on interface 0
Ethernet II, Src: cc:00:2b:08:00:00 (cc:00:2b:08:00:00), Dst: cc:01:2b:08:00:00 (cc
Internet Protocol Version 4, Src: 12.1.1.1 (12.1.1.1), Dst: 12.1.1.2 (12.1.1.2)
Transmission Control Protocol, Src Port: 14530 (14530), Dst Port: telnet (23), Seq:
    Source port: 14530 (14530)
    Destination port: telnet (23)
    [Stream index: 0]
    Sequence number: 71    (relative sequence number)
    Acknowledgment number: 1879    (relative ack number)
    Header length: 20 bytes
    Flags: 0x010 (ACK)
        000. .... .... = Reserved: Not set
        ...0 .... .... = Nonce: Not set
        .... 0... .... = Congestion Window Reduced (CWR): Not set
        .... .0.. .... = ECN-Echo: Not set
        .... ..0. .... = Urgent: Not set
        .... ...1 .... = Acknowledgment: Set
        .... .... 0... = Push: Not set
        .... .... .0.. = Reset: Not set
        .... .... ..0. = Syn: Not set
        .... .... ...0 = Fin: Not set
```

图 3-35　三次握手的第三个 TCP 分组

可以看到，此时客户端也向服务端返回确认位。

②以下验证 TCP 的可靠传输功能和 TELNET 的明文传输机制，单击 TELNET 分组，直到有明文数据出现，如图 3-36 所示。

```
Source        Destination    Protocol    Info
cc:01:2b:08CDP/VTP/DTPCDP    Device ID: R2    Port ID: FastEthernet0/0
12.1.1.1      12.1.1.2       TELNET     Telnet Data ...
12.1.1.2      12.1.1.1       TELNET     Telnet Data ...
12.1.1.2      12.1.1.1       TELNET     Telnet Data ...
12.1.1.2      12.1.1.1       TELNET     Telnet Data ...
12.1.1.2      12.1.1.1       TELNET     Telnet Data ...
12.1.1.2      12.1.1.1       TELNET     Telnet Data ...
Frame 19: 60 bytes on wire (480 bits), 60 bytes captured (480 bits) on interface 0
Ethernet II, Src: cc:00:2b:08:00:00 (cc:00:2b:08:00:00), Dst: cc:01:2b:08:00:00 (cc:01:2b:08:00:00)
Internet Protocol Version 4, Src: 12.1.1.1 (12.1.1.1), Dst: 12.1.1.2 (12.1.1.2)
Transmission Control Protocol, Src Port: 14530 (14530), Dst Port: telnet (23), Seq: 25, Ack: 67, Len: 1
Telnet
    Data: c
```

图 3-36　出现明文数据

此处可以看到，R1 向 R2 发送字符，所以 TELNET 协议中有字符【c】出现，序列号为 25，确认号为 67，长度为 1，单击下面的分组，如图 3-37 所示。

```
Source        Destination    Protocol    Info
cc:01:2b:08CDP/VTP/DTPCDP    Device ID: R2    Port ID: FastEthernet0/0
12.1.1.1      12.1.1.2       TELNET     Telnet Data ...
12.1.1.2      12.1.1.1       TELNET     Telnet Data ...
12.1.1.1      12.1.1.2       TELNET     Telnet Data ...
12.1.1.2      12.1.1.1       TELNET     Telnet Data ...
12.1.1.2      12.1.1.1       TELNET     Telnet Data ...
12.1.1.1      12.1.1.2       TELNET     Telnet Data ...
12.1.1.2      12.1.1.1       TELNET     Telnet Data ...
Frame 20: 60 bytes on wire (480 bits), 60 bytes captured (480 bits) on interface 0
Ethernet II, Src: cc:01:2b:08:00:00 (cc:01:2b:08:00:00), Dst: cc:00:2b:08:00:00 (cc:00:2b:08:00:00)
Internet Protocol Version 4, Src: 12.1.1.2 (12.1.1.2), Dst: 12.1.1.1 (12.1.1.1)
Transmission Control Protocol, Src Port: telnet (23), Dst Port: 14530 (14530), Seq: 67, Ack: 26, Len: 1
Telnet
    Data: c
```

图 3-37

此处 R2 对 R1 上次的数据包进行确认，序列号为 67，确认号为 26，长度为 1，单击下面的分组，如图 3-38 所示。

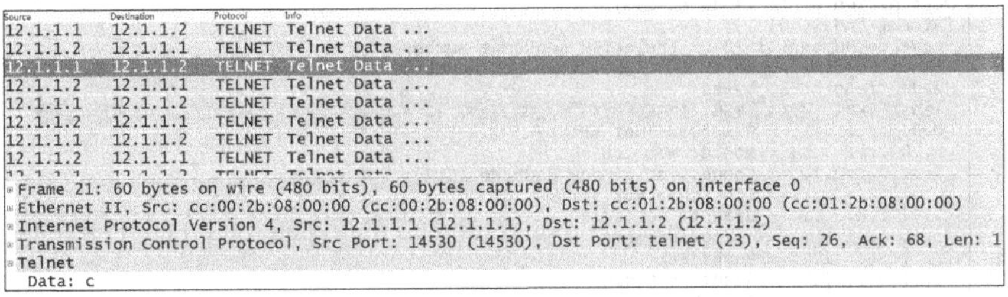

图 3-38

此处 R1 再次向 R2 发送【c】字符，序列号为 26，确认号为 68，长度为 1，单击下面分组，如图 3-39 所示。

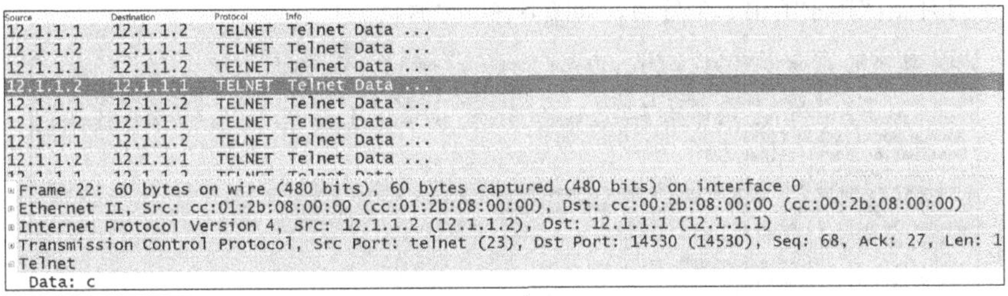

图 3-39

R2 再次对 R1 进行确认，序列号为 68，序列号为 27，长度为 1，单击下面分组，如图 3-40 所示。

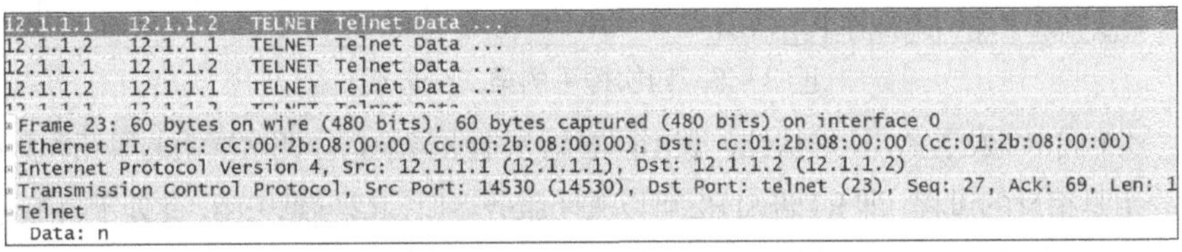

图 3-40

此时 R1 向 R2 发送字符【n】，序列号为 27，确认号为 69，长度为 1，以此类推……我们可以看到，本次接收者的确认号总是等于上次接收者的序列号加上数据包长度，即 Ack(n)=Seq(n-1)+Len，此为可靠传输下的结果。若数据包丢弃，则 Ack(n)<Seq(n-1)+Len，由此我们可以验证 TCP 的可靠更新机制。另外，在整个过程中，TELNET 远程登录的信息全部可以通过抓包查看，包括密码。所以在网络安全要求比较高的情况下，建议采用 SSH 协议来代替 TELNET 协议。

③以下验证 TCP 的四次挥手过程，如图 3-41 所示。

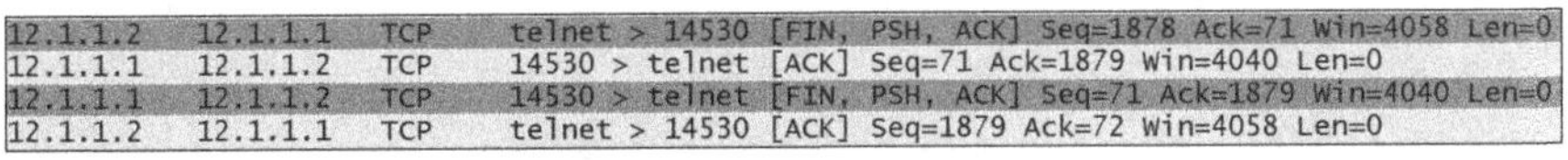

图 3-41 四次挥手过程

打开四次挥手的第一个分组，如图 3-42 所示。

第3章 TCP/IP 协议栈

```
Frame 120: 60 bytes on wire (480 bits), 60 bytes captured (480 bits) on interface 0
Ethernet II, Src: cc:01:2b:08:00:00 (cc:01:2b:08:00:00), Dst: cc:00:2b:08:00:00 (cc:
Internet Protocol Version 4, Src: 12.1.1.2 (12.1.1.2), Dst: 12.1.1.1 (12.1.1.1)
Transmission Control Protocol, Src Port: telnet (23), Dst Port: 14530 (14530), Seq:
  Source port: telnet (23)
  Destination port: 14530 (14530)
  [Stream index: 0]
  Sequence number: 1878    (relative sequence number)
  Acknowledgment number: 71    (relative ack number)
  Header length: 20 bytes
▶ Flags: 0x019 (FIN, PSH, ACK)
    000. .... .... = Reserved: Not set
    ...0 .... .... = Nonce: Not set
    .... 0... .... = Congestion Window Reduced (CWR): Not set
    .... .0.. .... = ECN-Echo: Not set
    .... ..0. .... = Urgent: Not set
    .... ...1 .... = Acknowledgment: Set
    .... .... 1... = Push: Set
    .... .... .0.. = Reset: Not set
    .... .... ..0. = Syn: Not set
    .... .... ...1 = Fin: Set
```

图 3-42　四次挥手的第一个分组

可以看到，此时 R2 向 R1 发送 FIN 结束位，告知对方可以关闭其服务端口。

打开四次挥手的第二个分组，如图 3-43 所示。

```
Frame 6: 60 bytes on wire (480 bits), 60 bytes captured (480 bits) on interface 0
Ethernet II, Src: cc:00:2b:08:00:00 (cc:00:2b:08:00:00), Dst: cc:01:2b:08:00:00 (cc:01:2b:08:00:00)
Internet Protocol Version 4, Src: 12.1.1.1 (12.1.1.1), Dst: 12.1.1.2 (12.1.1.2)
Transmission Control Protocol, Src Port: 14530 (14530), Dst Port: telnet (23), Seq: 1, Ack: 1, Len: 0
  Source port: 14530 (14530)
  Destination port: telnet (23)
  [Stream index: 0]
  Sequence number: 1    (relative sequence number)
  Acknowledgment number: 1    (relative ack number)
  Header length: 20 bytes
▶ Flags: 0x010 (ACK)
    000. .... .... = Reserved: Not set
    ...0 .... .... = Nonce: Not set
    .... 0... .... = Congestion Window Reduced (CWR): Not set
    .... .0.. .... = ECN-Echo: Not set
    .... ..0. .... = Urgent: Not set
    .... ...1 .... = Acknowledgment: Set
    .... .... 0... = Push: Not set
    .... .... .0.. = Reset: Not set
    .... .... ..0. = Syn: Not set
    .... .... ...0 = Fin: Not set
  Window size value: 4128
  [Calculated window size: 4128]
  [Window size scaling factor: -2 (no window scaling used)]
▶ Checksum: 0x383a [validation disabled]
```

图 3-43　四次挥手的第二个分组

R1 向 R2 发送 ACK 确认位，表示同意结束本服务端口。四次挥手的第三和第四个分组是相反的过程，即 R1 告知 R2 可以关闭服务端口，R2 同意关闭，此处不再抓包。通过 TCP 的四次挥手，可以关闭会话连接，节省设备资源消耗。此实验完成。

第 4 章 路由技术

本章主要学习几种路由技术，包括静态路由、RIP 路由协议、EIGRP 路由协议、OSPF 路由协议。路由技术是一门研究不同路由器之间如何寻路的课题，路由器通过路由协议的交互，可以相互学习到整个网络的拓扑，生成路由表，并为后续的数据转发提供基础的"导航"功能。以下为本章导航图：

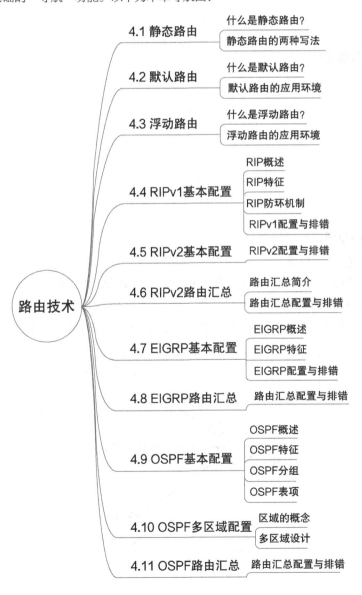

第 4 章 路由技术

4.1 静态路由

实验目的：
1. 掌握静态路由的基本配置。
2. 掌握静态路由两种写法的区别。

实验拓扑：

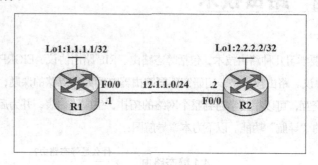

实验随手记：

_____
_____
_____
_____
_____

实验原理：

1. 静态路由概述

静态路由（Static Route）技术是路由技术发展的第一站，在这种技术背景下，网络管理员或工程师需要手工为路由器一条一条地录入路由条目，然后组成路由表。后续路由器根据这张手工的或者静态的路由表项，进行数据转发，如图 4-1 所示。

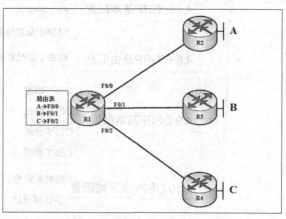

图 4-1 静态路由原理

【静态路由】
静态路由也被称为手工路由，严格意义上来讲，静态路由不属于路由协议，因为部署静态路由的路由器之间没有任何协议交互过程，即静态路由是本地有效的。

2. 静态路由特点

①一般运行在小型网络环境，无法支撑大型网络；

②无法动态适应网络拓扑变动，例如远端某条路径出现故障，本地路由不会自动删除而需要管理员手工介入删除；

③对路由器的系统资源占用非常小，并且不会占用网络带宽；

④由于编写语法非常简单，所以部署非常方便。

实验步骤：

1. 依据图中拓扑配置各设备的 IP 地址，并保证直连连通性。

在 R1 上进行配置，如下所示。

```
R1(config)#int f0/0
// 进入 f0/0 接口配置模式
R1(config-if)#no shutdown
// 打开该接口
R1(config-if)#ip address 12.1.1.1 255.255.255.0
// 配置 IP 地址信息
R1(config-if)#exit
// 退出接口模式
R1(config)#int loopback 1
//loopback 口称为环回口，可以模拟主机或网段，默认是打开的
R1(config-if)#ip address 1.1.1.1 255.255.255.255
// 为环回口配置 IP 地址
R1(config-if)#exit
```

在 R2 上进行置，如下所示。

```
R2(config)#int f0/0
R2(config-if)#no shutdown
R2(config-if)#ip address 12.1.1.2 255.255.255.0
R2(config-if)#exit
R2(config)#int loopback 1
R2(config-if)#ip address 2.2.2.2 255.255.255.255
R2(config-if)#exit
//R2 上同 R1 上一样，进行同样的配置
```

在 R1 上进行直连连通性测试，如下所示。

```
R1#ping 12.1.1.2
// 在 R1 上的 PING R2 的 f0/0 端口来测试两个路由器之间的连通性
Type escape sequence to abort.
```

【路由表】
用来存储路由条目的表项被称为路由表。路由表是由路由和路由所对应的输出接口组成的。一般情况下，路由表只存储到达目标的最佳路由，当有多条等价路由时，则执行负载均衡。

【环回接口】
一种逻辑接口，可以用来做链路测试，可以模拟一个网段或一台主机。路由器上对环回接口的数量是没有限制的。

【静态路由写法】
① ip route + 网段 + 掩码 + 下一跳地址；② ip route + 网段 + 掩码 + 本地输出接口。两种写法除了管理距离不同以外，在不同的网络环境也有讲究，例如广播网络环境一般采用下一跳地址，点对点网络环境一般采用本地出接口。从深入的层面来看，这涉及代理 ARP 和路由递归问题。

Sending 5, 100-byte ICMP Echos to 12.1.1.2, timeout is 2 seconds:
// 发送五个包，每个都是 100byte 的 ICMP 类型的包，目的地址是 12.1.1.2，超时时间为 2 秒

.!!!!
Success rate is 80 percent (4/5), round-trip min/avg/max = 28/36/48 ms

这说明直连连接没有问题。

2. 通过部署静态路由，使得 R1 和 R2 可以相互 Ping 通对方环回接口，如下所示。

R1(config)#ip route 2.2.2.2 255.255.255.255 12.1.1.2
// 写一条静态路由，指明通往目的 2.2.2.2/32 地址往 12.1.1.2 这个地址那里送

R2(config)#ip route 1.1.1.1 255.255.255.255 f0/0
// 写一条静态路由，指明通往目的 1.1.1.1/32 地址往 f0/0 这个端口送出去

查看路由表，R1 路由表如下所示。

R1#show ip route
// 该命令很常用，用来显示该路由器的路由表
Codes: C – connected, S – static, R – RIP, M – mobile, B – BGP
       D – EIGRP, EX – EIGRP external, O – OSPF, IA – OSPF inter area
       N1 – OSPF NSSA external type 1, N2 – OSPF NSSA external type 2
       E1 – OSPF external type 1, E2 – OSPF external type 2
       i – IS–IS, su – IS–IS summary, L1 – IS–IS level-1, L2 – IS–IS level-2
       ia – IS–IS inter area, * – candidate default, U – per-user static route
       o – ODR, P – periodic downloaded static route

Gateway of last resort is not set

     1.0.0.0/32 is subnetted, 1 subnets
C   1.1.1.1 is directly connected, Loopback1
//C 表示"直连"，即两个路由器是直接连接的
     2.0.0.0/32 is subnetted, 1 subnets
S   2.2.2.2 [1/0] via 12.1.1.2
// 采用下一跳 IP 写法得到的路由管理距离为 1，其中 S 代表"静态路由"
     12.0.0.0/24 is subnetted, 1 subnets
C   12.1.1.0 is directly connected, FastEthernet0/0

R2 路由表如下所示。

R2#show ip route

Codes: C – connected, S – static, R – RIP, M – mobile, B – BGP
       D – EIGRP, EX – EIGRP external, O – OSPF, IA – OSPF inter area

【路由代码】
路由类型代码是在路由条目之前用于标识此路由是通过哪种方式学到的。例如 C 表示直连路由，S 代表静态路由，R 表示 RIP 路由，O 表示 OSPF 路由，D 表示 EIGRP 路由等。

N1 – OSPF NSSA external type 1, N2 – OSPF NSSA external type 2
E1 – OSPF external type 1, E2 – OSPF external type 2
i – IS-IS, su – IS-IS summary, L1 – IS-IS level-1, L2 – IS-IS level-2
ia – IS-IS inter area, * – candidate default, U – per-user static route
o – ODR, P – periodic downloaded static route

Gateway of last resort is not set

  1.0.0.0/32 is subnetted, 1 subnets
S 1.1.1.1 is directly connected, FastEthernet0/0
// 特别注意，采用本地出接口的写法得到的路由管理距离为 0，与直连路由的管理距离一样

  2.0.0.0/32 is subnetted, 1 subnets
C 2.2.2.2 is directly connected, Loopback1

  12.0.0.0/24 is subnetted, 1 subnets
C 12.1.1.0 is directly connected, FastEthernet0/0

此时在 R1 和 R2 的路由表中生成"S"开头的静态路由，可以观察到两种静态路由写法得到的路由有差异，下一跳写法管理距离为 1，本地出接口写法管理距离为 0。

3. 在路由器上测试静态路由，如下所示。

R1#<u>ping 2.2.2.2 source 1.1.1.1</u>
// 以 1.1.1.1 这个地址为源地址，去 PING 2.2.2.2 这个地址
Type escape sequence to abort.
Sending 5, 100-byte ICMP Echos to 2.2.2.2, timeout is 2 seconds:
Packet sent with a source address of 1.1.1.1
!!!!!
// 成功
Success rate is 100 percent (5/5), round-trip min/avg/max = 28/38/52 ms

可以看出，通过部署静态路由全网连通。此实验完成。

【带源 Ping】
常规的 Ping 默认采用本地离目标最近的物理接口作为源地址，带源 Ping 可以修改与目标通信的源 IP 地址。

4.2 默认路由

实验目的:

1. 掌握默认路由的基本配置。
2. 理解默认路由的应用环境。

实验拓扑:

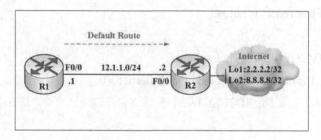

实验随手记:

_____
_____
_____
_____
_____

实验原理:

1. 默认路由概述

默认路由（Default Route）实际上是一种特殊的静态路由，它代表整张路由表的"最后一根稻草"，只有当本地路由表没法找到匹配的路由时，才会寻找默认路由进行数据转发，所以一般的默认路由放置在路由表的最底部。

2. 默认路由特点

默认路由一般部署在网络边缘或者互联网出口，例如校园网络的边缘路由器上，用于实现对互联网的访问。试想一下，整个互联网的路由条目有几十万条，如果采用静态路由技术来实现，网络管理员需要为路由器一条一条地录入这么多路由条目才能完整访问互联网，这该是一个多大的工程啊！况且，几十万条路由存储在路由表中，对设备本身的资源消耗就非常大，路由表搜寻延迟也比较长。

实验步骤:

1. 依据图中拓扑配置各设备的 IP 地址，并保证直连连通性。

在 R1 上进行配置，如下所示。

【默认路由】
企业网或校园网的出口连接到互联网运营商网络上，一般在这种情况下，网管人员需要在边缘路由器上部署默认路由。

```
R1(config)#int f0/0
// 进入 f0/0 接口配置模式
R1(config-if)#no shutdown
// 打开该接口
R1(config-if)#ip address 12.1.1.1 255.255.255.0
// 配置 IP 地址信息
R1(config-if)#exit
// 退出接口模式
R1(config)#int loopback 1
//loopback 口称为环回口，可以模拟主机或网段，默认是打开的
R1(config-if)#ip address 1.1.1.1 255.255.255.255
// 为环回口配置 IP 地址
R1(config-if)#exit
```

在 R2 上进行配置，如下所示。

```
R2(config)#int f0/0
R2(config-if)#no shutdown
R2(config-if)#ip address 12.1.1.2 255.255.255.0
R2(config-if)#exit
R2(config)#int loopback 1
R2(config-if)#ip address 2.2.2.2 255.255.255.255
// 上面部分都同 R1
R2(config-if)#exit
R2(config)#int loopback 2
// 再创建一个环回口
R2(config-if)#ip address 8.8.8.8 255.255.255.255
R2(config-if)#exit
```

在 R1 上进行直连连通性测试，如下所示。

```
R1#ping 12.1.1.2
// 在 R1 上面 PING R2 的 f0/0 端口，来测试两个路由器之间的连通性
Type escape sequence to abort.
Sending 5, 100-byte ICMP Echos to 12.1.1.2, timeout is 2 seconds:
.!!!!
Success rate is 80 percent (4/5), round-trip min/avg/max = 28/36/48 ms
```

这说明直连连接没有问题。

【全 0 IP】
0.0.0.0 在 IP 地址中代表全网。

2. 通过部署默认路由，R1 可以访问 R2 背后的所有环回网段，如下所示。

```
R1(config)#ip route 0.0.0.0 0.0.0.0 12.1.1.2
```
// 默认路由一般部署在互联网边缘，当路由表没有具体路由匹配目的时，则采用默认路由，所以默认路由一般处于路由表底部。也就是说，在路由表里面的路由都找不到之后，就只有用这一条路由发送出去了。

查看 R1 的路由表，具体如下所示。

```
R1#show ip route
// 该命令用来显示该路由的路由表
Codes: C – connected, S – static, R – RIP, M – mobile, B – BGP
       D – EIGRP, EX – EIGRP external, O – OSPF, IA – OSPF inter area
       N1 – OSPF NSSA external type 1, N2 – OSPF NSSA external type 2
       E1 – OSPF external type 1, E2 – OSPF external type 2
       i – IS–IS, su – IS–IS summary, L1 – IS–IS level–1, L2 – IS–IS level–2
       ia – IS–IS inter area, * – candidate default, U – per–user static route
       o – ODR, P – periodic downloaded static route

Gateway of last resort is 12.1.1.2 to network 0.0.0.0
// 如果数据包的目标网络不在路由表中，则将该数据包发往 12.1.1.2，默认路由条目中 0.0.0.0 是网络地址及子网掩码的通配符，表示任意网络

     1.0.0.0/32 is subnetted, 1 subnets
C       1.1.1.1 is directly connected, Loopback1
     12.0.0.0/24 is subnetted, 1 subnets
C       12.1.1.0 is directly connected, FastEthernet0/0
S*   0.0.0.0/0 [1/0] via 12.1.1.2
// 默认路由是一种特别的静态路由，比如要去 8.8.8.8 这个地址，但是路由表中都没有该路由条目，就只有匹配此路由，把它们全部往 12.1.1.2 这个地址发送
```

此时 R1 的路由表出现"S*"标识的路由条目，即静态默认路由。

3. 在 R1 上测试默认路由，如下所示。

```
R1#ping 2.2.2.2
Type escape sequence to abort.
Sending 5, 100-byte ICMP Echos to 2.2.2.2, timeout is 2 seconds:
!!!!!
// 成功
Success rate is 100 percent (5/5), round-trip min/avg/max = 28/38/52 ms
R1#ping 8.8.8.8
Type escape sequence to abort.
Sending 5, 100-byte ICMP Echos to 8.8.8.8, timeout is 2 seconds:
```

!!!!!
// 成功
Success rate is 100 percent (5/5), round-trip min/avg/max = 28/36/56 ms

以上表明，通过部署默认路由，R1 可以访问互联网的网段。此实验完成。

4.3 浮动路由

实验目的：
1. 掌握静态浮动路由的基本配置。
2. 理解静态浮动路由的主备切换。

实验拓扑：

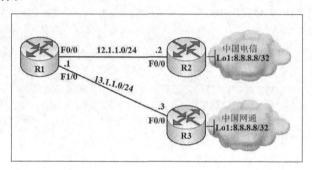

实验随手记：

_____
_____
_____
_____
_____

【浮动路由】
浮动路由一般出现在网络的多出口环境下，用于实现主备切换。浮动路由技术主要通过管理距离来实现。

实验原理：

1. 浮动路由概述

浮动路由（Floating Route）是在静态路由的基础上衍生过来的，可以实现多条静态路由的主备切换。企业网或者校园网一般有多条互联网出口链路，例如主链路租用中国电信专线，备用链路租用中国网通专线。在默认情况下，所有数据流往主链路走，当主链路出现故障之后，备用链路能够及时切换。

2. 浮动路由原理

浮动路由技术主要通过管理距离（优先级）来实现，通过将不同的静态路由设置不同的管理距离，路由器将最佳优先级的路由放入路由表，此时次优路由隐藏。当最优路由故障后，次优路由便"浮现"。

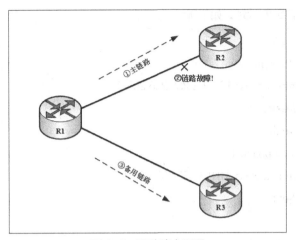

图 4-2 浮动路由原理

【管理距离】
从本质上来讲,管理距离是一种路由优先级,用于衡量路径的优劣,范围从 0 到 255,越小越优先。不同路由协议的默认管理距离不同,例如静态路由默认管理距离为 0 和 1,RIP 为 120,OSPF 为 110,EIGRP 为 90 等。可以通过命令对管理距离进行修改,实现路径控制。

实验步骤:

1. 依据图中拓扑配置各设备的 IP 地址,并保证直连连通性。

在 R1 上配置 IP 地址,如下所示。

R1(config)#int f0/0

// 进入接口模式

R1(config-if)#no shutdown

// 开启接口

R1(config-if)#ip address 12.1.1.1 255.255.255.0

// 配置 IP 地址

R1(config-if)#exit

// 退出接口模式

R1(config)#int f1/0

R1(config-if)#no shutdown

R1(config-if)#ip address 13.1.1.1 255.255.255.0

R1(config-if)#exit

// 原理同上

R1(config)#int loopback 1

// 创建环回接口

R1(config-if)#ip address 1.1.1.1 255.255.255.255

// 配置环回接口 IP 地址

R1(config-if)#exit

在 R2 上配置 IP 地址，如下所示。

```
R2(config)#int f0/0
R2(config-if)#no shutdown
R2(config-if)#ip address 12.1.1.2 255.255.255.0
R2(config-if)#exit
R2(config-if)#exit
R2(config)#int loopback 1
R2(config-if)#ip address 8.8.8.8 255.255.255.255
R2(config-if)#exit
// 原理同 R1
```

在 R3 上配置 IP 地址，如下所示。

```
R3(config)#int f0/0
R3(config-if)#no shutdown
R3(config-if)#ip address 13.1.1.3 255.255.255.0
R3(config-if)#exit
R3(config)#int loopback 1
R3(config-if)#ip address 8.8.8.8 255.255.255.255
R3(config-if)#exit
// 原理同 R1
```

在其中一台路由器上进行连通性测试，如下所示。

```
R1#ping 12.1.1.2
// 在 R1 上面 PING R2 的 f0/0 端口来测试两个路由器之间的连通性
Type escape sequence to abort.
Sending 5, 100-byte ICMP Echos to 12.1.1.2, timeout is 2 seconds:
// 发送五个包，每个都是 100byte 的 ICMP 类型的包，目的地址是 12.1.1.2，超时时间为 2 秒
.!!!!
Success rate is 80 percent (4/5), round-trip min/avg/max = 28/36/48 ms
R1#ping 13.1.1.3
Type escape sequence to abort.
Sending 5, 100-byte ICMP Echos to 13.1.1.3, timeout is 2 seconds:
.!!!!
// 成功
Success rate is 80 percent (4/5), round-trip min/avg/max = 16/31/48 ms
```

此时说明直连连接没有问题。

2. 在 R1 上部署浮动静态路由，如下所示。

```
R1(config)#ip route 0.0.0.0 0.0.0.0 12.1.1.2
R1(config)#ip route 0.0.0.0 0.0.0.0 13.1.1.3 100
// 默认情况下，默认路由的管理距离为 0，管理距离越小越优先，范围从 0 到
255。备用链路的管理距离要比主链路的管理距离大
```

查看 R1 的路由表，具体如下所示。

```
R1#show ip route
Codes: C – connected, S – static, R – RIP, M – mobile, B – BGP
       D – EIGRP, EX – EIGRP external, O – OSPF, IA – OSPF inter area
       N1 – OSPF NSSA external type 1, N2 – OSPF NSSA external type 2
       E1 – OSPF external type 1, E2 – OSPF external type 2
       i – IS-IS, su – IS-IS summary, L1 – IS-IS level-1, L2 – IS-IS level-2
       ia – IS-IS inter area, * – candidate default, U – per-user static route
       o – ODR, P – periodic downloaded static route

Gateway of last resort is 12.1.1.2 to network 0.0.0.0
// 如果数据包的目标网络不在路由表中，则将该数据包发往 12.1.1.2，默认路由条
目中 0.0.0.0 是网络地址及子网掩码的通配符，表示任意网络。
     1.0.0.0/32 is subnetted, 1 subnet
C       1.1.1.1 is directly connected, Loopback1
     12.0.0.0/24 is subnetted, 1 subnets
C       12.1.1.0 is directly connected, FastEthernet0/0
     13.0.0.0/24 is subnetted, 1 subnets
C       13.1.1.0 is directly connected, FastEthernet1/0
S*      0.0.0.0/0 [1/0] via 12.1.1.2
// 默认路由其实是一种特殊的静态路由，比如要去 8.8.8.8 这个地址，但是路由表
中都没有该路由条目，则只有匹配默认路由，并将数据往 12.1.1.2 发送。
```

由于管理距离的原因，备用路由被"隐藏"起来，只有最优的路由被放置进路由表中。

3. 此时进行浮动路由测试。

在 R2 和 R3 上开启 ICMP 实时调试，如下所示。

```
R2#debug ip icmp
// 打开 ICMP 报文调试，系统默认关闭 ICMP 报文调试信息
ICMP packet debugging is on
R3#debug ip icmp
ICMP packet debugging is on
```

此时在 R1 上 ping 互联网地址，如下所示。

【全 0 IP】
0.0.0.0 在 IP 地址中代表全网。

【Debug 调试】
与 show 命令不同，show 所看到的是某个时间点"静止"的信息，而 debug 命令所看到的信息是随着时间变动的动态进程输出过程，可以非常清晰地看到网络链路或拓扑的情况。debug 命令资源消耗非常大，一般实验室用的比较多，工程环境下慎用！

```
R1#ping 8.8.8.8
Type escape sequence to abort.
Sending 5, 100-byte ICMP Echos to 8.8.8.8, timeout is 2 seconds:
!!!!!
Success rate is 100 percent (5/5), round-trip min/avg/max = 28/36/48 ms
```

再到 R2 上查看，具体如下所示。

```
R2#debug ip icmp
ICMP packet debugging is on
*Mar  1 11:42:30.189: ICMP: echo reply sent, src 8.8.8.8, dst 12.1.1.1
// 回复 ICMP 的报文，源地址为 8.8.8.8，目的地址为 12.1.1.1
*Mar  1 11:42:30.237: ICMP: echo reply sent, src 8.8.8.8, dst 12.1.1.1
*Mar  1 11:42:30.269: ICMP: echo reply sent, src 8.8.8.8, dst 12.1.1.1
*Mar  1 11:42:30.317: ICMP: echo reply sent, src 8.8.8.8, dst 12.1.1.1
*Mar  1 11:42:30.345: ICMP: echo reply sent, src 8.8.8.8, dst 12.1.1.1
```

可以看到，R2 进行 ICMP 回应，说明正常情况下通往互联网的数据包往主链路走。此时关闭 R1 的主链路，模拟主链路出现故障，并查看路由表，如下所示。

```
R1(config)#int f0/0
R1(config-if)#shutdown
// 关闭 R1 的 f0/0 端口，让数据包不能从 R1 发送到 R2
R1(config-if)#end
R1#show ip route
// 再次查看路由表
Codes: C - connected, S - static, R - RIP, M - mobile, B - BGP
       D - EIGRP, EX - EIGRP external, O - OSPF, IA - OSPF inter area
       N1 - OSPF NSSA external type 1, N2 - OSPF NSSA external type 2
       E1 - OSPF external type 1, E2 - OSPF external type 2
       i - IS-IS, su - IS-IS summary, L1 - IS-IS level-1, L2 - IS-IS level-2
       ia - IS-IS inter area, * - candidate default, U - per-user static route
       o - ODR, P - periodic downloaded static route

Gateway of last resort is 13.1.1.3 to network 0.0.0.0

     1.0.0.0/32 is subnetted, 1 subnets
C       1.1.1.1 is directly connected, Loopback1
     13.0.0.0/24 is subnetted, 1 subnets
C       13.1.1.0 is directly connected, FastEthernet1/0
S*   0.0.0.0/0 [100/0] via 13.1.1.3
```

// 现在的网关变成了 13.1.1.3，即在路由表里面没有路由信息的数据包，直接往 R3 那里发送了

此时主链路路由消失，备用路由"浮出"路由表，再次在 R1 上 Ping 互联网地址，如下所示。

R1#ping 8.8.8.8

Type escape sequence to abort.
Sending 5, 100-byte ICMP Echos to 8.8.8.8, timeout is 2 seconds:
.!!!!
Success rate is 80 percent (4/5), round-trip min/avg/max = 24/30/32 ms

切换到 R3 上，如下所示。

R3#
*Mar 1 07:58:25.386: ICMP: echo reply sent, src 8.8.8.8, dst 13.1.1.1
// 从 R3 给 R1 的 ICMP 回复
*Mar 1 07:58:25.426: ICMP: echo reply sent, src 8.8.8.8, dst 13.1.1.1
*Mar 1 07:58:25.458: ICMP: echo reply sent, src 8.8.8.8, dst 13.1.1.1
*Mar 1 07:58:25.490: ICMP: echo reply sent, src 8.8.8.8, dst 13.1.1.1

从整个过程可以看到，当主链路正常时，备用链路隐藏，流量往主链路走；当主链路故障时，备用链路"浮出"，流量往备链路走，实现了故障切换。此实验完成。

4.4 RIPv1 基本配置

实验目的：

1. 掌握 RIPv1 的基本配置。
2. 掌握 RIPv1 的有类特性。

实验拓扑：

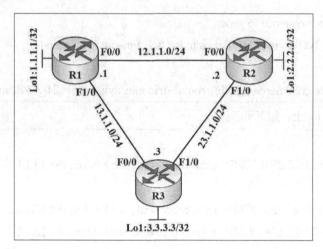

实验随手记：

_____
_____
_____
_____
_____
_____

实验原理：

1. RIP 概述

RIP（Routing Information Protocol，路由信息协议）是网络技术中真正最早的动态路由协议，同时，RIP 协议也是距离矢量协议的代表，它的年代离我们已经非常久远，是由施乐公司 Xerox 在 20 世纪 70 年代主导开发的，RIP 是路由技术发展的里程碑，是从静态路由到动态路由技术的变革点。目前，RIP 路由协议有三个版本，分别为 RIPv1、RIPv2、RIPng 版本，其中 RIPng 是在 IPv6 环境下运行的，其他 v1 和 v2 版本在 IPv4 下运行，本书只学习 v1 和 v2 版本。RIP 路由协议在当前的实际工程环境中已经不再使用，只是更多地在实验室环境下出现。通过学习 RIP 协议，我们可以更好地理解动态路由协议的特征。

2. RIP 特征

RIPv1 和 RIPv2 在特征上有相同点和不同点，如下所示。

【动态路由协议】
相比静态路由，动态路由协议能够适应在更加大型的网络中，能动态协商动态适应拓扑。常见的动态路由协议有 RIP、EIGRP、OSPF、ISIS 以及 BGP 协议。

【RIP 协议】
RIP 协议虽然在工程环境下不再使用，但是学习 RIP 可以帮助我们更好地理解距离矢量，更好地对比链路状态协议，它是动态路由协议的真正开端。

4.4 RIPv1 基本配置

协议	RIPv1	RIPv2
网络层次	应用层（基于 UDP520 端口）	应用层（基于 UDP520 端口）
路由算法	Bellman 算法	Bellman 算法
路由类型	有类（不支持 VLSM 和 CIDR）	无类（支持 VLSM 和 CIDR）
管理距离	120	120
度量值	跳数	跳数
更新方式	广播更新（255.255.255.255）	组播更新（224.0.0.9）
分组类型	Request 和 Response 分组	Request 和 Response 分组
路由汇总	不支持	支持

3. RIP 防环机制

水平分割、最大 16 跳、路由中毒、毒性逆转、抑制计时器。

4. RIP 的计时器

周期更新计时器(30s)、失效计时器(180s)、刷新计时器(180s)、抑制计时器(240s)。

实验步骤：

1. 依据图中拓扑配置各设备的 IP 地址，并保证直连连通性。

在 R1 上配置 IP 地址，如下所示。

```
R1(config)#int f0/0
// 进入接口模式
R1(config-if)#no shutdown
// 打开接口
R1(config-if)#ip address 12.1.1.1 255.255.255.0
// 配置 IP 地址
R1(config-if)#exit
// 退出接口模式
R1(config)#int f1/0
R1(config-if)#no shutdown
R1(config-if)#ip address 13.1.1.1 255.255.255.0
R1(config-if)#exit
// 原理同上
R1(config)#int loopback 1
// 创建环回接口
R1(config-if)#ip address 1.1.1.1 255.255.255.255
// 配置环回接口 IP 地址
R1(config-if)#exit
```

在 R2 上配置 IP 地址，如下所示。

【Bellman 算法】
贝尔曼算法是 RIP 的路由算法，用来计算 RIP 路由。

【有类和无类】
有类即 classful，是以前路由协议的特征，所有的路由条目遵循 A/B/C 地址分类方式，所有学到的路由必须是 /8 /16 /24 的掩码。有类环境下，路由协议不支持子网划分后的路由；而无类即 classless，是现在大部分路由协议的标准，不再遵循 A/B/C 标准，并且支持有掩码的路由条目。

【度量值】
用于衡量同一路由协议内部的路径优劣，一般越小越优先。

【跳数】
RIP 采用跳数作为其度量值，每台路由器为一跳，最大跳数为 16 跳，用来表明此路由失效。

```
R2(config)#int f0/0
R2(config-if)#no shutdown
R2(config-if)#ip address 12.1.1.2 255.255.255.0
R2(config-if)#exit
R2(config)#int f1/0
R2(config-if)#no shutdown
R2(config-if)#ip address 23.1.1.2 255.255.255.0
R2(config-if)#exit
R2(config)#int loopback 1
R2(config-if)#ip address 2.2.2.2 255.255.255.255
R2(config-if)#exit
// 同 R1
```

在 R3 上配置 IP 地址，如下所示。

```
R3(config)#int f0/0
R3(config-if)#no shutdown
R3(config-if)#ip address 13.1.1.3 255.255.255.0
R3(config-if)#exit
R3(config-if)#int f1/0
R3(config-if)#no shutdown
R3(config-if)#ip address 23.1.1.3 255.255.255.0
R3(config-if)#exit
R3(config)#int loopback 1
R3(config-if)#ip address 3.3.3.3 255.255.255.255
R3(config-if)#exit
// 同 R1
```

在其中一台路由器上进行连通性测试，如下所示。

```
R1#ping 12.1.1.2
// 在 R1 上面 PING R2 的 f0/0 端口来测试两个路由器之间的连通性
Type escape sequence to abort.

Sending 5, 100-byte ICMP Echos to 12.1.1.2, timeout is 2 seconds:
.!!!!

Success rate is 80 percent (4/5), round-trip min/avg/max = 28/36/48 ms


R1#ping 13.1.1.3
Type escape sequence to abort.
```

```
Sending 5, 100-byte ICMP Echos to 13.1.1.3, timeout is 2 seconds:
.!!!!
// 成功
Success rate is 80 percent (4/5), round-trip min/avg/max = 16/31/48 ms
```

此时说明直连连接没有问题。

2. 在每台路由器开始进行 RIPv1 的配置，R1 的配置如下所示。

```
R1(config)#router rip
// 配置模式下，进入 RIP 进程，默认的是 RIP V1 版本
R1(config-router)#network 12.0.0.0
// 注意，此处 network 是用于宣告主类网络号，不需要具体的路由条目如
12.1.1.0，也不需要路由掩码。
R1(config-router)#network 13.0.0.0
R1(config-router)#network 1.0.0.0
R1(config-router)#exit
```

R2 的配置如下所示。

```
R2(config)#router rip
R2(config-router)#network 12.0.0.0
R2(config-router)#network 23.0.0.0
R2(config-router)#network 2.0.0.0
R2(config-router)#exit
```

R3 的配置如下所示。

```
R3(config)#router rip
R3(config-router)#network 13.0.0.0
R3(config-router)#network 23.0.0.0
R3(config-router)#network 3.0.0.0
R3(config-router)#exit
```

3. 在任意一台路由器上查看路由表并进行测试。

在 R1 上查看路由表，具体如下所示。

```
R1#show ip route
// 查看路由表
Codes: C - connected, S - static, R - RIP, M - mobile, B - BGP
       D - EIGRP, EX - EIGRP external, O - OSPF, IA - OSPF inter area
       N1 - OSPF NSSA external type 1, N2 - OSPF NSSA external type 2
       E1 - OSPF external type 1, E2 - OSPF external type 2
       i - IS-IS, su - IS-IS summary, L1 - IS-IS level-1, L2 - IS-IS level-2
       ia - IS-IS inter area, * - candidate default, U - per-user static route
```

```
           o – ODR, P – periodic downloaded static route

Gateway of last resort is not set

       1.0.0.0/32 is subnetted, 1 subnets
C        1.1.1.1 is directly connected, Loopback1
R        2.0.0.0/8 [120/1] via 12.1.1.2, 00:00:01, FastEthernet0/0
// 由于 RIP 是有类的路由协议，并且 2.2.2.2 是 A 类地址，RIPv1 将其自动汇总。
此处 2.2.2.2/32 自动汇总为 2.0.0.0/8，3.3.3.3/32 则自动汇总为 3.0.0.0/8
R        3.0.0.0/8 [120/1] via 13.1.1.3, 00:00:07, FastEthernet1/0
R        23.0.0.0/8 [120/1] via 13.1.1.3, 00:00:07, FastEthernet1/0
                   [120/1] via 12.1.1.2, 00:00:01, FastEthernet0/0
       12.0.0.0/24 is subnetted, 1 subnets
C        12.1.1.0 is directly connected, FastEthernet0/0
C        13.1.1.0 is directly connected, FastEthernet1/0
```

此时看到路由表中出现"R"开头的路由条目，则为 R1 从 R2 和 R3 学习到的 RIPv1 路由，并且进行连通性测试，如下所示。

```
R1#ping 2.2.2.2 source 1.1.1.1
// 带源 PING 通，表示对方路由器也能 PING 通本地环回接口
Type escape sequence to abort.
Sending 5, 100-byte ICMP Echos to 2.2.2.2, timeout is 2 seconds:
Packet sent with a source address of 1.1.1.1
!!!!!
// 成功
Success rate is 100 percent (5/5), round-trip min/avg/max = 20/24/44 ms
R1#ping 3.3.3.3 source 1.1.1.1

Type escape sequence to abort.
Sending 5, 100-byte ICMP Echos to 3.3.3.3, timeout is 2 seconds:
Packet sent with a source address of 1.1.1.1
!!!!!
// 成功
Success rate is 100 percent (5/5), round-trip min/avg/max = 16/24/40 ms
```

在 R1 上测试成功，同样可以在 R2 和 R3 上进行测试。此实验完成。

4.5 RIPv2 基本配置

实验目的：
1. 掌握 RIPv2 的基本配置。
2. 掌握 RIPv2 的无类特性。

实验拓扑：

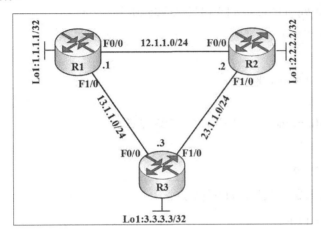

实验随手记：
_____
_____
_____
_____
_____
_____

实验原理：

RIPv2 是 RIPv1 的改进版本，主要加入了无类特征，例如子网掩码和路由汇总的支持。相比 RIPv1 来讲，RIPv2 能够更好地描述网络拓扑，实现更精准的路由条目传递。目前，RIPv2 更多地 在实验室环境下出现。

实验步骤：

1. 依据图中拓扑配置各设备的 IP 地址，并保证直连连通性。

在 R1 上配置 IP 地址，如下所示。

```
R1(config)#int f0/0
// 进入接口模式
R1(config-if)#no shutdown
// 打开接口
```

```
R1(config-if)#ip address 12.1.1.1 255.255.255.0
// 配置 IP 地址
R1(config-if)#exit
// 退出接口模式
R1(config)#int f1/0
R1(config-if)#no shutdown
R1(config-if)#ip address 13.1.1.1 255.255.255.0
R1(config-if)#exit
// 原理同上
R1(config)#int loopback 1
// 创建环回口
R1(config-if)#ip address 1.1.1.1 255.255.255.255
// 配置环回口 IP 地址
R1(config-if)#exit
```

在 R2 上配置 IP 地址，如下所示。

```
R2(config)#int f0/0
R2(config-if)#no shutdown
R2(config-if)#ip address 12.1.1.2 255.255.255.0
R2(config-if)#exit
R2(config)#int f1/0
R2(config-if)#no shutdown
R2(config-if)#ip address 23.1.1.2 255.255.255.0
R2(config-if)#exit
R2(config)#int loopback 1
R2(config-if)#ip address 2.2.2.2 255.255.255.255
R2(config-if)#exit
// 同 R1
```

在 R3 上配置 IP 地址，如下所示。

```
R3(config)#int f0/0
R3(config-if)#no shutdown
R3(config-if)#ip address 13.1.1.3 255.255.255.0
R3(config-if)#exit
R3(config-if)#int f1/0
R3(config-if)#no shutdown
R3(config-if)#ip address 23.1.1.3 255.255.255.0
R3(config-if)#exit
R3(config)#int loopback 1
```

4.5 RIPv2 基本配置

```
R3(config-if)#ip address 3.3.3.3 255.255.255.255
R3(config-if)#exit
// 同 R1
```

在其中一台路由器上进行连通性测试，如下所示。

```
R1#ping 12.1.1.2
// 在 R1 上面 PING R2 的 f0/0 端口来测试两个路由器之间的连通性
Type escape sequence to abort.

Sending 5, 100-byte ICMP Echos to 12.1.1.2, timeout is 2 seconds:
.!!!!
Success rate is 80 percent (4/5), round-trip min/avg/max = 28/36/48 ms


R1#ping 13.1.1.3
Type escape sequence to abort.
Sending 5, 100-byte ICMP Echos to 13.1.1.3, timeout is 2 seconds:
.!!!!
// 成功
Success rate is 80 percent (4/5), round-trip min/avg/max = 16/31/48 ms
```

此时说明直连连接没有问题。

2. 在每台路由器开始进行 RIPv2 的配置，R1 的配置如下所示。

```
R1(config)#router rip
// 开启 RIP 协议
R1(config-router)#version 2
// RIPv2 是在 RFC 1388 定义的，是对 RIPv1 的优化和补充
R1(config-router)#no auto-summary
// RIPv2 具备无类特性，支持 VLSM 和 CIDR；但默认情况下，RIPv2 开启自动汇
总，此命令用于关闭自动汇总特性
R1(config-router)#network 12.0.0.0
// 虽然 RIPv2 是无类路由协议，但是宣告路由的方式与 RIPv1 是一致的，还是宣
告主类网络号
R1(config-router)#network 13.0.0.0
R1(config-router)#network 1.0.0.0
R1(config-router)#exit
```

R2 的配置如下所示。

```
R2(config)#router rip
R2(config-router)#version 2
R2(config-router)#no auto-summary
```

// 原理同 R1

```
R2(config-router)#network 12.0.0.0
R2(config-router)#network 23.0.0.0
R2(config-router)#network 2.0.0.0
R2(config-router)#exit
```

R3 的配置如下所示。

```
R3(config)#router rip
R3(config-router)#version 2
R3(config-router)#no auto-summary
R3(config-router)#network 13.0.0.0
R3(config-router)#network 23.0.0.0
R3(config-router)#network 3.0.0.0
R3(config-router)#exit
```

3. 在任意一台路由器上查看路由表并进行测试。

在 R1 上查看路由表，具体如下所示。

```
R1#show ip route rip
```
// 此命令用于查看 RIP 路由；

同样的方法可以在命令 show ip route 后面加入关键词 connected、static、ospf、eigrp 直接看到具体协议的路由

```
        2.0.0.0/32 is subnetted, 1 subnets
R       2.2.2.2 [120/1] via 12.1.1.2, 00:00:04, FastEthernet0/0
```
// RIPv2 路由更新夹带具体的掩码信息，相比 RIPv1 能更好地描述网络

```
        3.0.0.0/32 is subnetted, 1 subnets
R       3.3.3.3 [120/1] via 13.1.1.3, 00:00:12, FastEthernet1/0
        23.0.0.0/24 is subnetted, 1 subnets
R       23.1.1.0 [120/1] via 13.1.1.3, 00:00:12, FastEthernet1/0
                 [120/1] via 12.1.1.2, 00:00:04, FastEthernet0/0
```

可以看到 R1 从 R2 和 R3 学习到的 RIPv2 路由，此时进行连通性测试，如下所示。

```
R1#ping 2.2.2.2 source 1.1.1.1
```
// 带源 PING 通，表示对方路由器也能 PING 通本地环回接口

```
Type escape sequence to abort.
Sending 5, 100-byte ICMP Echos to 2.2.2.2, timeout is 2 seconds:
Packet sent with a source address of 1.1.1.1
!!!!!
```
// 成功

```
Success rate is 100 percent (5/5), round-trip min/avg/max = 20/24/44 ms
R1#ping 3.3.3.3 source 1.1.1.1
```

```
Type escape sequence to abort.
Sending 5, 100-byte ICMP Echos to 3.3.3.3, timeout is 2 seconds:
Packet sent with a source address of 1.1.1.1
!!!!!
Success rate is 100 percent (5/5), round-trip min/avg/max = 16/24/40 ms
```

此时在 R1 上测试成功，同样方法可以在 R2 和 R3 上也测试成功，表明通过 RIPv2 三台路由器的环回网段都相互连通。此实验完成。

4.6 RIPv2 路由汇总

实验目的:

掌握 RIPv2 的路由汇总。

实验拓扑:

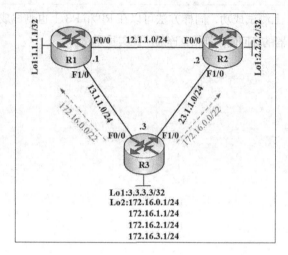

实验随手记:

___

【路由汇总】
路由汇总技术是将多条路由条目"压缩"成一条或若干条路由条目,可以有效节省路由条目。

实验原理:

路由汇总是路由网络中非常常见的路由优化技术,路由协议如 RIPv2 通过执行路由汇总,将多条精细路由汇总成一条路由并通告出去。这样做的好处有很多,例如,路由器的路由表体积变小,数据搜寻所花费的时间减少;另外,路由更新信息对网络的带宽资源占用也变小。路由汇总一般部署在网络的边界,例如分支站点需要将分支的路由通告给总部时,通常将分支的精细路由进行汇总后,再发送给总部站点。

实验步骤:

1. 依据图中拓扑配置各设备的 IP 地址,并保证直连连通性。

在 R1 上配置 IP 地址,如下所示。

```
R1(config)#int f0/0
// 进入接口
R1(config-if)#no shutdown
```

```
// 打开接口
R1(config-if)#ip address 12.1.1.1 255.255.255.0
// 配置接口 IP
R1(config-if)#exit
// 退出接口模式
R1(config)#int f1/0
R1(config-if)#no shutdown
R1(config-if)#ip address 13.1.1.1 255.255.255.0
R1(config-if)#exit
// 原理同上
R1(config)#int loopback 1
// 创建环回口
R1(config-if)#ip address 1.1.1.1 255.255.255.255
// 为环回口配置 IP 地址
R1(config-if)#exit
```

在 R2 上配置 IP 地址，如下所示。

```
R2(config)#int f0/0
R2(config-if)#no shutdown
R2(config-if)#ip address 12.1.1.2 255.255.255.0
R2(config-if)#exit
R2(config)#int f1/0
R2(config-if)#no shutdown
R2(config-if)#ip address 23.1.1.2 255.255.255.0
R2(config-if)#exit
R2(config)#int loopback 1
R2(config-if)#ip address 2.2.2.2 255.255.255.255
R2(config-if)#exit
```

在 R3 上配置 IP 地址，如下所示。

```
R3(config)#int f0/0
R3(config-if)#no shutdown
R3(config-if)#ip address 13.1.1.3 255.255.255.0
R3(config-if)#exit
R3(config-if)#int f1/0
R3(config-if)#no shutdown
R3(config-if)#ip address 23.1.1.3 255.255.255.0
R3(config-if)#exit
R3(config)#int loopback 1
```

【从地址】
从地址是用于实现同一接口多个IP的解决方案，一般在实验室环境下用于测试。

```
R3(config-if)#ip address 3.3.3.3 255.255.255.255
R3(config-if)#exit
R3(config)#int loopback 2
R3(config-if)#ip address 172.16.0.1 255.255.255.0
R3(config-if)#ip address 172.16.1.1 255.255.255.0 secondary
// Secondary 表示从地址或者辅助地址，通过此参数可以在同一接口下配置多个IP地址，并简化配置。这是为了实验方便才如此配置。
R3(config-if)#ip address 172.16.2.1 255.255.255.0 secondary
R3(config-if)#ip address 172.16.3.1 255.255.255.0 secondary
R3(config-if)#exit
```

在其中一台路由器上进行连通性测试，如下所示。

```
R1#ping 12.1.1.2
// 在 R1 上面 PING R2 的 f0/0 端口来测试两个路由器之间的连通性
Type escape sequence to abort.
// 这句话的意思是"输入转义序列以终止"，其实，就是输入"Shift+Ctrl+6"来终止该命令。PING 命令默认发送五个包，所以无需特别干预。但假如发的包很多时，则可以派上用场
Sending 5, 100-byte ICMP Echos to 12.1.1.2, timeout is 2 seconds:
.!!!!
Success rate is 80 percent (4/5), round-trip min/avg/max = 28/36/48 ms
```

```
R1#ping 13.1.1.3
Type escape sequence to abort.
Sending 5, 100-byte ICMP Echos to 13.1.1.3, timeout is 2 seconds:
.!!!!
Success rate is 80 percent (4/5), round-trip min/avg/max = 16/31/48 ms
```

此时说明直连连接没有问题。

2. 在每台路由器开始进行 RIPv2 的配置，R1 的配置如下所示。

```
R1(config)#router rip
// 开启 RIP 协议进程
R1(config-router)#version 2
// 把 RIP 设置为版本 2
R1(config-router)#no auto-summary
// RIPv2 具备无类特性，支持 VLSM 和 CIDR；但默认情况下，RIPv2 开启自动汇总，此命令用于关闭自动汇总特性【标配】
R1(config-router)#network 12.0.0.0
R1(config-router)#network 13.0.0.0
```

```
R1(config-router)#network 1.0.0.0
R1(config-router)#exit
```

R2 的配置如下所示。

```
R2(config)#router rip
R2(config-router)#version 2
R2(config-router)#no auto-summary
R2(config-router)#network 12.0.0.0
R2(config-router)#network 23.0.0.0
R2(config-router)#network 2.0.0.0
R2(config-router)#exit
```

R3 的配置如下所示。

```
R3(config)#router rip
R3(config-router)#version 2
R3(config-router)#no auto-summary
R3(config-router)#network 13.0.0.0
R3(config-router)#network 23.0.0.0
R3(config-router)#network 3.0.0.0
R3(config-router)#network 172.16.0.0
// 宣告主类网络号，此时将 172.16.0.0 所在子网路由全部通告出去
R3(config-router)#exit
```

在 R1 和 R2 上查看路由表，其中 R1 上具体如下所示。

```
R1#show ip route rip
// 此命令用于查看 RIP 路由；
同样的方法可以在命令 show ip route 后面加入关键词 connected、static、ospf、
eigrp 直接看到具体协议的路由
     2.0.0.0/32 is subnetted, 1 subnets
R       2.2.2.2 [120/1] via 12.1.1.2, 00:00:04, FastEthernet0/0
// RIPv2 路由更新夹带具体的掩码信息，相比 RIPv1 能更好地描述网络
     3.0.0.0/32 is subnetted, 1 subnets
R       3.3.3.3 [120/1] via 13.1.1.3, 00:00:27, FastEthernet1/0
     23.0.0.0/24 is subnetted, 1 subnets
R       23.1.1.0 [120/1] via 13.1.1.3, 00:00:27, FastEthernet1/0
                 [120/1] via 12.1.1.2, 00:00:19, FastEthernet0/0
     172.16.0.0/24 is subnetted, 4 subnets
R       172.16.0.0 [120/1] via 13.1.1.3, 00:00:27, FastEthernet1/0
R       172.16.1.0 [120/1] via 13.1.1.3, 00:00:27, FastEthernet1/0
R       172.16.2.0 [120/1] via 13.1.1.3, 00:00:27, FastEthernet1/0
```

R   172.16.3.0 [120/1] via 13.1.1.3, 00:00:27, FastEthernet1/0
// 四条精细的路由条目被宣告进来了

R2 上具体如下所示。

```
R2#show ip route rip
     1.0.0.0/32 is subnetted, 1 subnets
R       1.1.1.1 [120/1] via 12.1.1.1, 00:00:10, FastEthernet0/0
     3.0.0.0/32 is subnetted, 1 subnets
R       3.3.3.3 [120/1] via 23.1.1.3, 00:00:14, FastEthernet1/0
     172.16.0.0/24 is subnetted, 4 subnets
R       172.16.0.0 [120/1] via 23.1.1.3, 00:00:14, FastEthernet1/0
R       172.16.1.0 [120/1] via 23.1.1.3, 00:00:14, FastEthernet1/0
R       172.16.2.0 [120/1] via 23.1.1.3, 00:00:14, FastEthernet1/0
R       172.16.3.0 [120/1] via 23.1.1.3, 00:00:14, FastEthernet1/0
     13.0.0.0/24 is subnetted, 1 subnets
R       13.1.1.0 [120/1] via 23.1.1.3, 00:00:14, FastEthernet1/0
```

可以看到 R1 和 R2 从 R3 上学到 4 条精细子网。

3. 为了缩小路由表，提高路由器转发效率，在 R3 上执行路由汇总，配置如下所示。

```
R3(config)#int f0/0
R3(config-if)#ip summary-address rip 172.16.0.0 255.255.252.0
//172.16.0.0 255.255.252.0 为 4 条精细网段的汇总网段
R3(config-if)#exit
R3(config)#int f1/0
R3(config-if)#ip summary-address rip 172.16.0.0 255.255.252.0
// RIPv2 路由汇总在接口下执行，并且需要在多个接口同时执行，否则精细路由会从其他接口"溜走"
R3(config-if)#exit
```

再到 R1 和 R2 上查看路由表，如下所示。

```
R1#show ip route rip
     2.0.0.0/24 is subnetted, 1 subnets
R       2.2.2.0 [120/1] via 12.1.1.2, 00:00:00, FastEthernet0/0
     3.0.0.0/32 is subnetted, 1 subnets
R       3.3.3.3 [120/1] via 13.1.1.3, 00:00:01, FastEthernet1/0
     23.0.0.0/24 is subnetted, 1 subnets
R       23.1.1.0 [120/1] via 13.1.1.3, 00:00:01, FastEthernet1/0
                [120/1] via 12.1.1.2, 00:00:00, FastEthernet0/0
     172.16.0.0/22 is subnetted, 1 subnets
```

// 被汇总了，注意子网掩码变成了 /22 了！
R   172.16.0.0 [120/1] via 13.1.1.3, 00:00:01, FastEthernet1/0

R2#show ip route rip
       1.0.0.0/32 is subnetted, 1 subnets
R   1.1.1.1 [120/1] via 12.1.1.1, 00:00:18, FastEthernet0/0
       3.0.0.0/32 is subnetted, 1 subnets
R   3.3.3.3 [120/1] via 23.1.1.3, 00:00:13, FastEthernet1/0
       172.16.0.0/22 is subnetted, 1 subnets
// 在 R2 这里同样被汇总了
R   172.16.0.0 [120/1] via 23.1.1.3, 00:00:13, FastEthernet1/0
       13.0.0.0/24 is subnetted, 1 subnets
R   13.1.1.0 [120/1] via 23.1.1.3, 00:00:13, FastEthernet1/0
                  [120/1] via 12.1.1.1, 00:00:18, FastEthernet0/0

可以看到，原来的 4 条精细路由已经变成 1 条汇总路由，路由表的体积变小，达到了路由汇总的目的。此实验完成。

4.7 EIGRP 基本配置

实验目的：
1. 掌握 EIGRP 的基本配置。
2. 掌握 EIGRP 的邻居表、拓扑表、路由表。
3. 掌握 EIGRP 的无类特性。

实验拓扑：

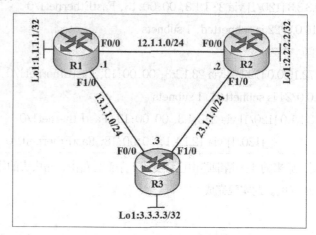

实验随手记：

_____
_____
_____
_____
_____
_____

实验原理：

1. EIGRP 概述

EIGRP（Enhanced Interior Gateway Routing Protocol，增强型内部网关路由协议）是由思科公司自主开发的路由协议，相比 RIP 协议，EIGRP 有更多优越的路由特性。EIGRP 采用 DUAL 路由算法进行路径选择，能够对网络拓扑的变动做出迅速的反应。EIGRP 结合了距离矢量协议和链路状态协议的特点，可以将其称为高级距离矢量协议或者混合路由协议。

2. EIGRP 特点

特征	描述
网络层次	传输层（基于 IP 协议号 88）
路由算法	DUAL(弥散更新算法)

【EIGRP 协议】
EIGRP 的前身是 IGRP，EIGRP 在原有的基础上加入了很多新的特性，例如无类特性。EIGRP 在性能上能够满足大型网络，相比 RIP 而言，其拓展性大大增强。

续表

特征	描述
协议类型	无类路由协议（支持 VLSM 和 CIDR）
协议特征	高级距离矢量协议
管理距离	90 和 170
度量值	混合度量值（带宽、延迟、负载、可信度、MTU）
更新方式	触发更新、组播更新（224.0.0.10）、增量更新、可靠更新

3. EIGRP 分组

EIGRP 有多种路由分组，常见分组有 Hello 分组、Update 分组、Ack 分组、Query 分组、Reply 分组，各自功能如下所示。

分组	功能
Hello 分组	用于建立和维持邻居关系
Update 分组	用于承载和传递路由条目
Ack 分组	用于实现 EIGRP 的可靠传输
Query 分组	用于当路由丢失时，向邻居发送路由请求
Reply 分组	用于收到路由请求时，给予回复

4. EIGRP 表项

EIGRP 有三张表，分别为邻居表、拓扑表、路由表，功能如下。

邻居表：用于存储邻居关系。

拓扑表：用于存储所有路由条目。

路由表：用于存储最佳路由条目。

实验步骤：

1. 依据图中拓扑配置各设备的 IP 地址，并保证直连连通性。

在 R1 上配置 IP 地址，如下所示。

```
R1(config)#int f0/0
// 进入接口模式
R1(config-if)#no shutdown
// 打开接口
R1(config-if)#ip address 12.1.1.1 255.255.255.0
// 配置 IP 地址
R1(config-if)#exit
// 退出接口模式
R1(config)#int f1/0
R1(config-if)#no shutdown
R1(config-if)#ip address 13.1.1.1 255.255.255.0
R1(config-if)#exit
// 原理同上
```

【DUAL 算法】
DUAL 算法是 EIGRP 的路由算法，用来计算 EIGRP 的最佳路径。

【高级距离矢量协议】
EIGRP 不是单纯的距离矢量协议，它具备链路状态协议的一些特征，所以有时候也被称为高级距离矢量协议或者混合路由协议。

【触发更新】
表示网络有变化，则马上发送更新信息通告邻居，可以更好地响应网络变动。

【组播更新】
即采用目的地址为组播地址，而不是广播地址。

【增量更新】
即 EIGRP 在拓扑变动过程中，只发送有改变的路由条目信息，而不会像 RIP 一样，发送整张路由表。

```
R1(config)#int loopback 1
// 创建环回接口
R1(config-if)#ip address 1.1.1.1 255.255.255.255
// 配置环回 IP 地址
R1(config-if)#exit
```

在 R2 上配置 IP 地址,如下所示。

```
R2(config)#int f0/0
R2(config-if)#no shutdown
R2(config-if)#ip address 12.1.1.2 255.255.255.0
R2(config-if)#exit
R2(config)#int f1/0
R2(config-if)#no shutdown
R2(config-if)#ip address 23.1.1.2 255.255.255.0
R2(config-if)#exit
R2(config)#int loopback 1
R2(config-if)#ip address 2.2.2.2 255.255.255.255
R2(config-if)#exit
```

在 R3 上配置 IP 地址,如下所示。

```
R3(config)#int f0/0
R3(config-if)#no shutdown
R3(config-if)#ip address 13.1.1.3 255.255.255.0
R3(config-if)#exit
R3(config-if)#int f1/0
R3(config-if)#no shutdown
R3(config-if)#ip address 23.1.1.3 255.255.255.0
R3(config-if)#exit
R3(config)#int loopback 1
R3(config-if)#ip address 3.3.3.3 255.255.255.255
R3(config-if)#exit
```

在其中一台路由器上进行连通性测试,如下所示。

```
R1#ping 12.1.1.2
// 在 R1 上面 PING R2 的 f0/0 端口来测试两个路由器之间的连通性
Type escape sequence to abort.

Sending 5, 100-byte ICMP Echos to 12.1.1.2, timeout is 2 seconds:
.!!!!
Success rate is 80 percent (4/5), round-trip min/avg/max = 28/36/48 ms
```

```
R1#ping 13.1.1.3
Type escape sequence to abort.
Sending 5, 100-byte ICMP Echos to 13.1.1.3, timeout is 2 seconds:
.!!!!
Success rate is 80 percent (4/5), round-trip min/avg/max = 16/31/48 ms
```

此时说明直连连接没有问题。

2. 在每台路由器开始进行 EIGRP 的配置，R1 的配置如下所示。

```
R1(config)#router eigrp 100
// 进入 EIGRP 进程，并且定义此路由器所在自治系统号为 100
R1(config-router)#no auto-summary
// 默认情况下，EIGRP 开启自动汇总，通过此命令关闭自动汇总
R1(config-router)#network 12.1.1.0 0.0.0.255
// network + 网段 + 反掩码，用来宣告路由信息
R1(config-router)#network 13.1.1.0 0.0.0.255
R1(config-router)#network 1.1.1.1 0.0.0.0
R1(config-router)#exit
//EIGRP 宣告路由的方式有三种，如下所示。
network + 主类网络号。
network + 网段 + 子网掩码。
network + 网段 + 反掩码。
在本例中，通告路由条目 12.1.1.0/24，则可以通过 network 12.0.0.0 或 network
12.1.1.0 255.255.255.0 或者 network 12.1.1.0 0.0.0.255 来实现
```

R2 的配置如下所示。

```
R2(config)#router eigrp 100
R2(config-router)#no auto-summary
R2(config-router)#network 12.1.1.0 0.0.0.255
R2(config-router)#network 23.1.1.0 0.0.0.255
R2(config-router)#network 2.2.2.2 0.0.0.0
R2(config-router)#exit
```

R3 的配置如下所示。

```
R3(config)#router eigrp 100
R3(config-router)#no auto-summary
R3(config-router)#network 13.1.1.0 0.0.0.255
R3(config-router)#network 23.1.1.0 0.0.0.255
R3(config-router)#network 3.3.3.3 0.0.0.0
R3(config-router)#exit
```

【自治系统】
AS 号自治系统号，范围从 1 到 65535，代表路由管理域。在 EIGRP 协议中，不同自治系统之间的路由相互隔离，不可通信。

【自动汇总】
这是有类路由协议如 RIPv1 和 IGRP 的特性，即所有交互的路由条目，A 类路由的掩码都为 /8，B 类为 /16，C 类为 /24，所有路由条目都被汇总为主类。自动汇总没法准确地传递路由条目，如一条 1.1.1.1/32 的路由会被汇总为 1.0.0.0/8。所以在 RIPv2 和 EIGRP 中都需要关闭自动汇总。

3. 查看 EIGRP 的邻居表、拓扑表和路由表，在 R1 上查看 EIGRP 邻居表，如下所示。

```
R1#show ip eigrp neighbors
// 此命令可以用来查看 EIGRP 的邻居表
IP-EIGRP neighbors for process 100
H    Address    Interface   Hold Uptime    SRTT    RTO    Q      Seq
                            (sec)          (ms)           Cnt    Num
1    13.1.1.3   Fa1/0       11   00:03:45  74      444    0      11
0    12.1.1.2   Fa0/0       14   00:04:10  65      390    0      10
// H 栏—按照发现顺序列出邻居
//Address —该邻居的 IP 地址
//Interface —收到此 Hello 数据包的本地接口
//Hold —邻居保持时间，每次收到 Hello 数据包时，此值即被重置为最大保留时间，然后倒计时，到零为止，如果到达了零，则认为该邻居进入失效
//Uptime（运行时间）—从该邻居被添加到邻居表以来的时间
//SRTT（平均回程计时器）和 RTO（重传间隔）—由 RTP 用于管理可靠 EIGRP 数据包
//Queue Count（队列数）—接口队列延迟，一般为零，如果大于零，则说明有 EIGRP 数据包等待发送
//Sequence Number（序列号）—用于跟踪更新、查询和应答数据包
```

其中 H 表示与其他路由器建立邻居的顺序；Address 表示邻居的 IP 地址；Interface 表示本地接口；Hold 表示 holdtime，默认为 15s。从这里可以看到 R1 与 R2 和 R3 依次建立了邻居关系。

在 R1 上查看 EIGRP 的拓扑表，如下所示。

```
R1#show ip eigrp topology
// 查看 EIGRP 的拓扑结构
IP-EIGRP Topology Table for AS(100)/ID(1.1.1.1)

Codes: P – Passive, A – Active, U – Update, Q – Query, R – Reply,
       r – reply Status, s – sia Status

P 3.3.3.3/32, 1 successors, FD is 156160
        via 13.1.1.3 (156160/128256), FastEthernet1/0
P 2.2.2.2/32, 1 successors, FD is 156160
        via 12.1.1.2 (156160/128256), FastEthernet0/0
P 1.1.1.1/32, 1 successors, FD is 128256
        via Connected, Loopback1
P 12.1.1.0/24, 1 successors, FD is 28160
```

```
        via Connected, FastEthernet0/0
P 13.1.1.0/24, 1 successors, FD is 28160
        via Connected, FastEthernet1/0
P 23.1.1.0/24, 2 successors, FD is 30720
        via 13.1.1.3 (30720/28160), FastEthernet1/0
        via 12.1.1.2 (30720/28160), FastEthernet0/0
```

此命令放置满足 FC<可行条件>的路由条目,若想查看所有 EIGRP 的路由条目,则需要加入 all-links 参数,如下所示。

```
R1#show ip eigrp topology all-links
// 查看 EIGRP 全部链路信息
IP-EIGRP Topology Table for AS(100)/ID(1.1.1.1)

Codes: P – Passive, A – Active, U – Update, Q – Query, R – Reply,
       r – reply Status, s – sia Status
P 3.3.3.3/32, 1 successors, FD is 156160, serno 8
        via 13.1.1.3 (156160/128256), FastEthernet1/0
        via 12.1.1.2 (158720/156160), FastEthernet0/0
P 2.2.2.2/32, 1 successors, FD is 156160, serno 5
        via 12.1.1.2 (156160/128256), FastEthernet0/0
        via 13.1.1.3 (158720/156160), FastEthernet1/0
P 1.1.1.1/32, 1 successors, FD is 128256, serno 3
        via Connected, Loopback1
P 12.1.1.0/24, 1 successors, FD is 28160, serno 1
        via Connected, FastEthernet0/0
P 13.1.1.0/24, 1 successors, FD is 28160, serno 2
        via Connected, FastEthernet1/0
P 23.1.1.0/24, 2 successors, FD is 30720, serno 7
        via 13.1.1.3 (30720/28160), FastEthernet1/0
        via 12.1.1.2 (30720/28160), FastEthernet0/0
```

此时查看 R1 上的路由表,如下所示。

```
R1#show ip route eigrp
// 查看路由表中 EIGRP 协议的相关路由条目。
        2.0.0.0/32 is subnetted, 1 subnets
D       2.2.2.2 [90/156160] via 12.1.1.2, 00:08:48, FastEthernet0/0
//EIGRP 是一种无类路由协议,支持 VLSM 和 CIDR,路由更新夹带掩码信息
        3.0.0.0/32 is subnetted, 1 subnets
D       3.3.3.3 [90/156160] via 13.1.1.3, 00:08:47, FastEthernet1/0
```

```
         23.0.0.0/24 is subnetted, 1 subnets
D        23.1.1.0 [90/30720] via 13.1.1.3, 00:08:50, FastEthernet1/0
                  [90/30720] via 12.1.1.2, 00:08:50, FastEthernet0/0
```

可以看到，R1 已经通过 EIGRP 协议从其他路由器学习到 EIGRP 路由，此时可以进行测试，如下所示。

```
R1#ping 2.2.2.2 source 1.1.1.1
Type escape sequence to abort.
Sending 5, 100-byte ICMP Echos to 2.2.2.2, timeout is 2 seconds:
Packet sent with a source address of 1.1.1.1
!!!!!
// 成功
Success rate is 100 percent (5/5), round-trip min/avg/max = 24/33/48 ms


R1#ping 3.3.3.3 source 1.1.1.1
Type escape sequence to abort.
Sending 5, 100-byte ICMP Echos to 3.3.3.3, timeout is 2 seconds:
Packet sent with a source address of 1.1.1.1
!!!!!
// 成功
Success rate is 100 percent (5/5), round-trip min/avg/max = 32/37/48 ms
```

此时，R1 可以与 R2 和 R3 的环回网段进行通信；同样的方法在 R2 和 R3 上进行测试。此实验完成。

4.8　EIGRP 路由汇总

实验目的：
1. 掌握 EIGRP 的路由汇总。
2. 理解 EIGRP 汇总时的 null 路由。

实验拓扑：

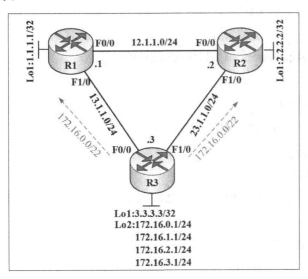

实验随手记：

_____
_____
_____
_____
_____
_____

实验步骤：
1. 依据图中拓扑配置各设备的 IP 地址，并保证直连连通性。
   在 R1 上配置 IP 地址，如下所示。

R1(config)#int f0/0
// 进入接口模式
R1(config-if)#no shutdown
// 打开接口
R1(config-if)#ip address 12.1.1.1 255.255.255.0
// 配置 IP 地址
R1(config-if)#exit

```
// 退出接口模式
R1(config)#int f1/0
R1(config-if)#no shutdown
R1(config-if)#ip address 13.1.1.1 255.255.255.0
R1(config-if)#exit
// 原理同上
R1(config)#int loopback 1
// 创建环回接口
R1(config-if)#ip address 1.1.1.1 255.255.255.255
// 配置环回 IP 地址
R1(config-if)#exit
```

在 R2 上配置 IP 地址，如下所示。

```
R2(config)#int f0/0
R2(config-if)#no shutdown
R2(config-if)#ip address 12.1.1.2 255.255.255.0
R2(config-if)#exit
R2(config)#int f1/0
R2(config-if)#no shutdown
R2(config-if)#ip address 23.1.1.2 255.255.255.0
R2(config-if)#exit
R2(config)#int loopback 1
R2(config-if)#ip address 2.2.2.2 255.255.255.255
R2(config-if)#exit
```

在 R3 上配置 IP 地址，如下所示。

```
R3(config)#int f0/0
R3(config-if)#no shutdown
R3(config-if)#ip address 13.1.1.3 255.255.255.0
R3(config-if)#exit
R3(config-if)#int f1/0
R3(config-if)#no shutdown
R3(config-if)#ip address 23.1.1.3 255.255.255.0
R3(config-if)#exit
R3(config)#int loopback 1
R3(config-if)#ip address 3.3.3.3 255.255.255.255
R3(config-if)#exit
R3(config)#int loopback 2
R3(config-if)#ip address 172.16.0.1 255.255.255.0
```

R3(config-if)#ip address 172.16.1.1 255.255.255.0 secondary

R3(config-if)#ip address 172.16.2.1 255.255.255.0 secondary

R3(config-if)#ip address 172.16.3.1 255.255.255.0 secondary

R3(config-if)#exit

在其中一台路由器上进行连通性测试，如下所示。

R1#ping 12.1.1.2

// 在 R1 上面 PING R2 的 f0/0 端口来测试两个路由器之间的连通性

Type escape sequence to abort.

Sending 5, 100-byte ICMP Echos to 12.1.1.2, timeout is 2 seconds:

.!!!!

Success rate is 80 percent (4/5), round-trip min/avg/max = 28/36/48 ms

R1#ping 13.1.1.3

Type escape sequence to abort.

Sending 5, 100-byte ICMP Echos to 13.1.1.3, timeout is 2 seconds:

.!!!!

Success rate is 80 percent (4/5), round-trip min/avg/max = 16/31/48 ms

此时说明直连连接没有问题。

2. 在每台路由器开始进行 EIGRP 的配置，R1 的配置如下所示。

  R1(config)#router eigrp 100

// 进入 EIGRP 进程，并且定义此路由器所在自治系统号为 100

  R1(config-router)#no auto-summary

// 关闭自动汇总

  R1(config-router)#network 12.1.1.0 0.0.0.255

// 默认情况下，EIGRP 开启自动汇总，通过此命令关闭自动汇总

  R1(config-router)#network 13.1.1.0 0.0.0.255

  R1(config-router)#network 1.1.1.1 0.0.0.0

  R1(config-router)#exit

//EIGRP 宣告路由的方式有三种，如下所示。

network + 主类网络号。

network + 网段 + 子网掩码。

network + 网段 + 反掩码。

在本例中，通告路由条目 12.1.1.0/24，则可以通过 network 12.0.0.0 或 network 12.1.1.0 255.255.255.0 或者 network 12.1.1.0 0.0.0.255 来实现

R2 的配置如下所示。

```
R2(config)#router eigrp 100
R2(config-router)#no auto-summary
R2(config-router)#network 12.1.1.0 0.0.0.255
R2(config-router)#network 23.1.1.0 0.0.0.255
R2(config-router)#network 2.2.2.2 0.0.0.0
R2(config-router)#exit
```

R3 的配置如下所示。

```
R3(config)#router eigrp 100
R3(config-router)#no auto-summary
R3(config-router)#network 13.1.1.0 0.0.0.255
R3(config-router)#network 23.1.1.0 0.0.0.255
R3(config-router)#network 3.3.3.3 0.0.0.0
R3(config-router)#network 172.16.0.0 0.0.255.255
// 把 172.16.0.0/16 这个网段宣告进去，可以将 4 个精细网段全部宣告进去
R3(config-router)#exit
```

3. 查看 R1 和 R2 的路由表，如下所示。

```
R1#show ip route eigrp
// 查看路由表
     2.0.0.0/32 is subnetted, 1 subnets
D       2.2.2.2 [90/156160] via 12.1.1.2, 00:08:48, FastEthernet0/0
//EIGRP 是一种无类路由协议，支持 VLSM 和 CIDR，路由更新夹带掩码信息
     3.0.0.0/32 is subnetted, 1 subnets
D       3.3.3.3 [90/156160] via 13.1.1.3, 00:41:16, FastEthernet1/0
     23.0.0.0/24 is subnetted, 1 subnets
D       23.1.1.0 [90/30720] via 13.1.1.3, 00:41:19, FastEthernet1/0
                 [90/30720] via 12.1.1.2, 00:41:19, FastEthernet0/0
     172.16.0.0/24 is subnetted, 4 subnets
D       172.16.0.0 [90/156160] via 13.1.1.3, 00:00:06, FastEthernet1/0
// 可以清楚地看到那 4 条路由信息被成功地宣告进来了，并被其他路由器学习到了
D       172.16.1.0 [90/156160] via 13.1.1.3, 00:00:06, FastEthernet1/0
D       172.16.2.0 [90/156160] via 13.1.1.3, 00:00:06, FastEthernet1/0
D       172.16.3.0 [90/156160] via 13.1.1.3, 00:00:06, FastEthernet1/0
```

```
R2#show ip route eigrp
     1.0.0.0/32 is subnetted, 1 subnets
```

```
D        1.1.1.1 [90/156160] via 12.1.1.1, 00:41:29, FastEthernet0/0
     3.0.0.0/32 is subnetted, 1 subnets
D        3.3.3.3 [90/156160] via 23.1.1.3, 00:41:28, FastEthernet1/0
     172.16.0.0/24 is subnetted, 4 subnets
D        172.16.0.0 [90/156160] via 23.1.1.3, 00:00:18, FastEthernet1/0
D        172.16.1.0 [90/156160] via 23.1.1.3, 00:00:18, FastEthernet1/0
D        172.16.2.0 [90/156160] via 23.1.1.3, 00:00:18, FastEthernet1/0
D        172.16.3.0 [90/156160] via 23.1.1.3, 00:00:18, FastEthernet1/0
     13.0.0.0/24 is subnetted, 1 subnets
D        13.1.1.0 [90/30720] via 23.1.1.3, 00:41:29, FastEthernet1/0
                  [90/30720] via 12.1.1.1, 00:41:29, FastEthernet0/0
```

可以看到，此时 R1 和 R2 从 R3 学到 4 条精细路由。为缩减路由表并提高路由器转发效率，将此 4 条精细子网进行汇总。

4. 在 R3 上执行路由汇总，如下所示。

```
R3(config)#int f0/0
// 在接口下配置
R3(config-if)#ip summary-address eigrp 100 172.16.0.0 255.255.252.0
// 注意，命令中要带 EIGRP 的自治系统号，其汇总路由掩码是 /22 位的
R3(config)#int f1/0
R3(config-if)#ip summary-address eigrp 100 172.16.0.0 255.255.252.0
// 为了防止从其他端口泄露，所以每个参与进程下的端口都需要配置
R3(config-if)#exit
```

在 R1 和 R2 上再次查看路由表，如下所示。

```
R1#show ip route eigrp
// 查看路由表中 EIGRP 协议的相关路由条目
     2.0.0.0/32 is subnetted, 1 subnets
D        2.2.2.2 [90/156160] via 12.1.1.2, 00:54:23, FastEthernet0/0
     3.0.0.0/32 is subnetted, 1 subnets
D        3.3.3.3 [90/156160] via 13.1.1.3, 00:54:23, FastEthernet1/0
     23.0.0.0/24 is subnetted, 1 subnets
D        23.1.1.0 [90/30720] via 13.1.1.3, 00:54:26, FastEthernet1/0
                  [90/30720] via 12.1.1.2, 00:54:26, FastEthernet0/0
     172.16.0.0/22 is subnetted, 1 subnets
D        172.16.0.0 [90/156160] via 13.1.1.3, 00:00:06, FastEthernet1/0
// 注意，此时汇总路由的子网掩码已经变成了 /22 了，说明汇总成功！
R2#show ip route eigrp
     1.0.0.0/32 is subnetted, 1 subnets
```

```
D        1.1.1.1 [90/156160] via 12.1.1.1, 00:54:29, FastEthernet0/0
     3.0.0.0/32 is subnetted, 1 subnets
D        3.3.3.3 [90/156160] via 23.1.1.3, 00:54:28, FastEthernet1/0
     172.16.0.0/22 is subnetted, 1 subnets
D        172.16.0.0 [90/156160] via 23.1.1.3, 00:00:13, FastEthernet1/0
     13.0.0.0/24 is subnetted, 1 subnets
D        13.1.1.0 [90/30720] via 23.1.1.3, 00:54:29, FastEthernet1/0
                 [90/30720] via 12.1.1.1, 00:54:29, FastEthernet0/0
```

此时可以看到，原本 4 条精细路由汇总成 1 条路由，说明汇总成功。在 R3 上查看路由表，如下所示。

```
R3#show ip route eigrp
     1.0.0.0/32 is subnetted, 1 subnets
D        1.1.1.1 [90/156160] via 13.1.1.1, 00:58:02, FastEthernet0/0
     2.0.0.0/32 is subnetted, 1 subnets
D        2.2.2.2 [90/156160] via 23.1.1.2, 00:58:02, FastEthernet1/0
     172.16.0.0/16 is variably subnetted, 5 subnets, 2 masks
D        172.16.0.0/22 is a summary, 00:03:47, Null0
     12.0.0.0/24 is subnetted, 1 subnets
D        12.1.1.0 [90/30720] via 23.1.1.2, 00:58:02, FastEthernet1/0
                  [90/30720] via 13.1.1.1, 00:58:02, FastEthernet0/0
```

上面的 NULL 路由是 EIGRP 汇总时自动生成的，是用于解决不精确汇总可能引发的路由环路。此实验完成。

4.9 OSPF 基本配置

实验目的：
1. 掌握 OSPF 的单区域配置。
2. 掌握 OSPF 的邻居表、数据库、路由表。
3. 理解 OSPF 的进程号、路由器标识、区域概念。

实验拓扑：

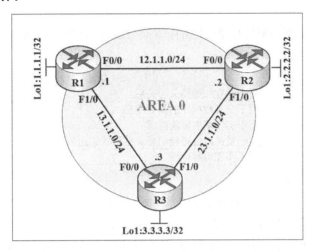

实验随手记：

_____
_____
_____
_____
_____

实验原理：

1. OSPF 概述

OSPF（Open Shortest Path First，开放最短路径优先）路由协议是目前使用最广泛的路由协议之一，是由 IETF 在 20 世纪 80 年代开发的。目前主要有两个 OSPF 版本，一个是 OSPFv2，应用在 IPv4 环境，一个是 OSPFv3，应用在 IPv6 环境，本书只介绍 OSPFv2。OSPF 路由协议是链路状态协议的代表，路由更新采用链路状态通告信息而不是简单的路由条目，每个路由器可以得知整个网络的拓扑结构，并且采用 SPF 进行路由计算，适应于大型网络拓扑环境，通过区域划分实现层次型拓扑结构。

【OSPF 协议】
OSPF 可以说是目前网络工程师最熟悉的路由协议，它的运行机制与 RIP 和 EIGRP 都有很大的差别，采用层次性网络设计，通过区域划分极大地拓展了网络的体积。除此之外，OSPF 是一种链路状态协议，它对网络拓扑了解一清二楚，可以绝对防止环路发生。

第 4 章 路由技术

【链路状态协议】
不同于距离矢量协议，链路状态协议交互的信息不是路由条目，而是链路信息。同时，距离矢量的路由条目是通过邻居学习到的，而链路状态协议是本地生成的。

【SPF 算法】
SPF 算法是 OSPF 的路由算法，用于计算以自己为中心到达目标的最短路径。

【开销】
开销 <cost> 是 OSPF 用于衡量路径优劣的参数。类似 RIP 的跳数，不过 OSPF 的开销值是基于带宽而不是基于路由器的个数，能够更好地反映网络拓扑。

2. OSPF 特点

特征	描述
网络层次	传输层（基于 IP 协议号 89）
路由算法	SPF 算法（最短路径优先算法）
协议类型	无类路由协议（支持 VLSM 和 CIDR）
协议特征	链路状态协议
管理距离	110
度量值	开销值（Cost=100Mbit/ 带宽）
更新方式	组播更新（224.0.0.5 和 224.0.0.6）

3. OSPF 分组

OSPF 有多种路由分组，常见分组有 Hello 分组、DBD 分组、LSR 分组、LSU 分组、LSAck 分组，各自功能如下所示。

分组	功能
Hello 分组	用于建立和维持邻居关系
DBD 分组	用于交互数据库的简要信息
LSR 分组	用于实现路由更新请求
LSU 分组	用于发送路由更新信息（链路状态通告信息）
LSAck 分组	用于实现可靠更新

4. OSPF 表项

OSPF 有三张表，分别为邻居表、数据库、路由表。

邻居表：用于存储邻居关系。

数据库：用于存储所有的链路状态信息（链路状态数据库相当于网络地图）。

路由表：用于存储最佳路由条目。

实验步骤：

1. 依据图中拓扑配置各设备的 IP 地址，并保证直连连通性。

在 R1 上配置 IP 地址，如下所示。

R1(config)#int f0/0

// 进入接口模式

R1(config-if)#no shutdown

// 打开接口

R1(config-if)#ip address 12.1.1.1 255.255.255.0

// 配置 IP 地址

R1(config-if)#exit

// 退出接口模式

R1(config)#int f1/0

R1(config-if)#no shutdown

R1(config-if)#ip address 13.1.1.1 255.255.255.0

R1(config-if)#exit

// 原理同上

R1(config)#int loopback 1

```
// 创建环回接口
R1(config-if)#ip address 1.1.1.1 255.255.255.255
// 配置环回 IP 地址
R1(config-if)#exit
```

在 R2 上配置 IP 地址，如下所示。

```
R2(config)#int f0/0
R2(config-if)#no shutdown
R2(config-if)#ip address 12.1.1.2 255.255.255.0
R2(config-if)#exit
R2(config)#int f1/0
R2(config-if)#no shutdown
R2(config-if)#ip address 23.1.1.2 255.255.255.0
R2(config-if)#exit
R2(config)#int loopback 1
R2(config-if)#ip address 2.2.2.2 255.255.255.255
R2(config-if)#exit
```

在 R3 上配置 IP 地址，如下所示。

```
R3(config)#int f0/0
R3(config-if)#no shutdown
R3(config-if)#ip address 13.1.1.3 255.255.255.0
R3(config-if)#exit
R3(config-if)#int f1/0
R3(config-if)#no shutdown
R3(config-if)#ip address 23.1.1.3 255.255.255.0
R3(config-if)#exit
R3(config)#int loopback 1
R3(config-if)#ip address 3.3.3.3 255.255.255.255
R3(config-if)#exit
```

在其中一台路由器上进行连通性测试，如下所示。

```
R1#ping 12.1.1.2
// 在 R1 上面 PING R2 的 f0/0 端口来测试两个路由器之间的连通性
Type escape sequence to abort.


Sending 5, 100-byte ICMP Echos to 12.1.1.2, timeout is 2 seconds:
.!!!!
```

```
Success rate is 80 percent (4/5), round-trip min/avg/max = 28/36/48 ms
R1#ping 13.1.1.3
Type escape sequence to abort.
Sending 5, 100-byte ICMP Echos to 13.1.1.3, timeout is 2 seconds:
.!!!!
Success rate is 80 percent (4/5), round-trip min/avg/max = 16/31/48 ms
```

此时说明直连连接没有问题。

2. 在每台路由器上开始进行 OSPF 的配置，R1 的配置如下所示。

```
R1(config)#router ospf 100
// Routerospf process-id,process-id 指的是 OSPF 的进程号，范围从 1 到 65535
R1(config-router)#router-id 1.1.1.1
// Router-id 指的是路由器标识，一般称为 RID，用于在网络中唯一标识一台路由器
R1(config-router)#network 12.1.1.0 0.0.0.255 area 0
//Network + 网段 + 反掩码 + 区域号来宣告路由信息
R1(config-router)#network 13.1.1.0 0.0.0.255 area 0
R1(config-router)#network 1.1.1.1 0.0.0.0 area 0
R1(config-router)#exit
```

//Route-id 即路由器标识，一般称为 RID，用于在网络中唯一标识一台路由器，可以手工指定，也可以由路由器自动选举，选举规则如下所示。

1. 手工指定 < 最优先 >。
2. 选择环回接口 IP 最大的。
3. 选举物理接口 IP 最大的。

一般环回接口不管是手工指定还是自动选举，都会采用环回接口的 IP 地址以保证稳定性。

R2 的配置如下所示。

```
R2(config)#router ospf 100
R2(config-router)#router-id 2.2.2.2
R2(config-router)#network 12.1.1.0 0.0.0.255 area 0
R2(config-router)#network 23.1.1.0 0.0.0.255 area 0
R2(config-router)#network 2.2.2.2 0.0.0.0 area 0
R2(config-router)#exit
```

R3 的配置如下所示。

```
R3(config)#router ospf 100
R3(config-router)#router-id 3.3.3.3
R3(config-router)#network 13.1.1.0 0.0.0.255 area 0
R3(config-router)#network 23.1.1.0 0.0.0.255 area 0
```

【进程号】
OSPF 进程号是本地有效的，用于标识本地 OSPF 进程的数值。路由器可以同时部署多个 OSPF 进程，不同 OSPF 进程之间的路由相互隔离。

【区域号】
区域号是 OSPF 用于实现区域划分的工具，通过划分区域，可以使网络从扁平化走向层次化，除了方便管理，还可以减少路由动荡，减轻路由器压力；区域号范围从 0 到 4294967295，区域分为骨干区域和常规区域，一般区域 0 表示骨干区域，用于连接其他常规区域并实现流量中转。

```
R3(config-router)#network 3.3.3.3 0.0.0.0 area 0
R3(config-router)#exit
```

3. 查看 OSPF 的邻居表、数据库以及路由表，例如在 R1 上查看邻居表，如下所示。

```
R1#show ip ospf neighbor
// 查看 OSPF 相关邻居信息
Neighbor ID     Pri   State       Dead Time     Address       Interface
3.3.3.3         1     FULL/DR     00:00:39      13.1.1.3      FastEthernet1/0
2.2.2.2         1     FULL/DR     00:00:31      12.1.1.2      FastEthernet0/0
```

其中 Neighbor ID 表示邻居的 RID，Pri 表示邻居的接口优先级，State 表示邻居与本地的状态以及邻居的 DR/BDR/Drother 角色。从这里可以看出，R1 与 R2 和 R3 已经建立邻接关系，并且 R2 和 R3 都是 DR。

在 R1 上查看 OSPF 链路状态数据库，如下所示。

```
R1#show ip ospf database
// 显示 OSPF 的数据库

            OSPF Router with ID (1.1.1.1) (Process ID 100)

                Router Link States (Area 0)

Link ID         ADV Router      Age         Seq#          Checksum Link count
1.1.1.1         1.1.1.1         277         0x8000000B    0x00625B 3
2.2.2.2         2.2.2.2         243         0x80000006    0x00544A 3
3.3.3.3         3.3.3.3         244         0x80000003    0x00B2DE 3


                Net Link States (Area 0)

Link ID         ADV Router      Age         Seq#          Checksum
12.1.1.1        1.1.1.1         277         0x80000003    0x0046D2
13.1.1.1        1.1.1.1         282         0x80000003    0x006BA8
23.1.1.3        3.3.3.3         243         0x80000001    0x00AE4F
```

OSPF 的数据库放置着整个 OSPF 区域的"地图"信息，即 LSA 信息。

在 R1 上查看 OSPF 路由表，如下所示。

```
R1#show ip route ospf
// 查看路由表中有关 OSPF 路由信息
      2.0.0.0/32 is subnetted, 1 subnets
O        2.2.2.2 [110/2] via 12.1.1.2, 00:07:31, FastEthernet0/0
// 路由信息是带子网掩码的，同时，我们可以看到路由的管理距离是 110
      3.0.0.0/32 is subnetted, 1 subnets
```

```
O        3.3.3.3 [110/2] via 13.1.1.3, 00:07:31, FastEthernet1/0
     23.0.0.0/24 is subnetted, 1 subnets
O        23.1.1.0 [110/2] via 13.1.1.3, 00:07:31, FastEthernet1/0
                  [110/2] via 12.1.1.2, 00:07:31, FastEthernet0/0
// 到 23.0.0.0/24 这个网段，有两条路径可以抵达，且它们的代价是一样的
```

此时可以看到，R1 从 R2 和 R3 上学到其环回网段，进行连通性测试，如下所示。

```
R1#ping 2.2.2.2 source 1.1.1.1
// 用 1.1.1.1 为源地址，去 PING 2.2.2.2
Type escape sequence to abort.
Sending 5, 100-byte ICMP Echos to 2.2.2.2, timeout is 2 seconds:
Packet sent with a source address of 1.1.1.1
!!!!!
// 成功
Success rate is 100 percent (5/5), round-trip min/avg/max = 16/26/40 ms
R1#ping 3.3.3.3 source 1.1.1.1

Type escape sequence to abort.
Sending 5, 100-byte ICMP Echos to 3.3.3.3, timeout is 2 seconds:
Packet sent with a source address of 1.1.1.1
!!!!!
// 成功
Success rate is 100 percent (5/5), round-trip min/avg/max = 20/24/44 ms
```

同样的方法在 R2 和 R3 上进行测试，可以发现，通过运行 OSPF 协议可以实现全网连通。此实验完成。

4.10 OSPF 多区域配置

实验目的：
1. 掌握 OSPF 的多区域配置。
2. 理解 OSPF 的路由类型 <O 和 O IA 路由 >。
3. 理解 OSPF 的路由器角色 < 常规路由器和边界路由器 >。

实验拓扑：

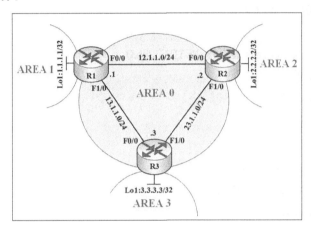

实验随手记：

_____
_____
_____
_____
_____

实验原理：

1. 区域概述

OSPF 的区域划分可以将整个网络切割成不同区域，以校园网为例，这样做的好处有很多：从管理层面来讲，将不同宿舍楼放置在不同的区域，再指派不同的网络管理员负责不同的区域，这样网络管理可以分域分工化；从技术角度来讲，多区域划分之后，不同区域的路由器只需要维护本地的数据库而不需要维护整网的数据库，对于系统资源消耗大大降低，另外，不同区域的网络动荡也在区域边界上被隔离，使得整个网络更加稳定。

2. 区域类型

OSPF 有多种区域类型，最常见的区域为骨干区域和常规区域，其中骨干区域一般以数值 AREA 0 出现，其他常规区域必须围绕在骨干区域周边。网络的核心一般放置在骨干区域，网络的分支点放置在常规区域。例如校园网中，所有宿舍楼可以放入

【区域】
区域 <AREA> 是 OSPF 协议用于划分和隔离网络拓扑的工具。不同区域之间的链路状态数据库相互隔离，但是可以通过边界路由器进行通信。区域划分可以有效减少路由器的资源消耗，并且方便网络运维人员对网络进行管理。

不同的常规区域，然后再连接到校园网的核心层即骨干区域。

实验步骤：

1. 依据图中拓扑配置各设备的 IP 地址，并保证直连连通性。

在 R1 上配置 IP 地址，如下所示。

```
R1(config)#int f0/0
// 进入接口模式
R1(config-if)#no shutdown
// 打开接口
R1(config-if)#ip address 12.1.1.1 255.255.255.0
// 配置 IP 地址
R1(config-if)#exit
// 退出接口模式
R1(config)#int f1/0
R1(config-if)#no shutdown
R1(config-if)#ip address 13.1.1.1 255.255.255.0
R1(config-if)#exit
// 原理同上
R1(config)#int loopback 1
// 创建环回接口
R1(config-if)#ip address 1.1.1.1 255.255.255.255
// 配置环回 IP 地址
R1(config-if)#exit
```

在 R2 上配置 IP 地址，如下所示。

```
R2(config)#int f0/0
R2(config-if)#no shutdown
R2(config-if)#ip address 12.1.1.2 255.255.255.0
R2(config-if)#exit
R2(config)#int f1/0
R2(config-if)#no shutdown
R2(config-if)#ip address 23.1.1.2 255.255.255.0
R2(config-if)#exit
R2(config)#int loopback 1
R2(config-if)#ip address 2.2.2.2 255.255.255.255
R2(config-if)#exit
```

在 R3 上配置 IP 地址，如下所示。

```
R3(config)#int f0/0
R3(config-if)#no shutdown
R3(config-if)#ip address 13.1.1.3 255.255.255.0
R3(config-if)#exit
R3(config-if)#int f1/0
R3(config-if)#no shutdown
R3(config-if)#ip address 23.1.1.3 255.255.255.0
R3(config-if)#exit
R3(config)#int loopback 1
R3(config-if)#ip address 3.3.3.3 255.255.255.255
R3(config-if)#exit
```

在其中一台路由器上进行连通性测试，如下所示。

```
R1#ping 12.1.1.2
// 在 R1 上面 PING R2 的 f0/0 端口来测试两个路由器之间的连通性
Type escape sequence to abort.
Sending 5, 100-byte ICMP Echos to 12.1.1.2, timeout is 2 seconds:
.!!!!
Success rate is 80 percent (4/5), round-trip min/avg/max = 28/36/48 ms


R1#ping 13.1.1.3
Type escape sequence to abort.
Sending 5, 100-byte ICMP Echos to 13.1.1.3, timeout is 2 seconds:
.!!!!
Success rate is 80 percent (4/5), round-trip min/avg/max = 16/31/48 ms
```

此时说明直连连接没有问题。

2. 在每台路由器上开始进行 OSPF 的配置，R1 的配置如下所示。

```
R1(config)#router ospf 100
// 在配置模式下开启 OSPF 路由协议，进程号是 100
R1(config-router)#router-id 1.1.1.1
// 设置 router-id 为 1.1.1.1，即用 1.1.1.1 这个地址来标识本路由器
R1(config-router)#network 12.1.1.0 0.0.0.255 area 0
// 用 Network 加 12.1.1.0 和反掩码（子网掩码取反）来宣告该条路由信息
R1(config-router)#network 13.1.1.0 0.0.0.255 area 0
R1(config-router)#network 1.1.1.1 0.0.0.0 area 1
// 特别注意，这个宣告时候设置的是 Area 1
R1(config-router)#exit
```

R2 的配置如下所示。

```
R2(config)#router ospf 100
R2(config-router)#router-id 2.2.2.2
R2(config-router)#network 12.1.1.0 0.0.0.255 area 0
R2(config-router)#network 23.1.1.0 0.0.0.255 area 0
R2(config-router)#network 2.2.2.2 0.0.0.0 area 2
// 特别注意，这个宣告时候设置的是 Area 2
R2(config-router)#exit
```

R3 的配置如下所示。

```
R3(config)#router ospf 100
R3(config-router)#router-id 3.3.3.3
R3(config-router)#network 13.1.1.0 0.0.0.255 area 0
R3(config-router)#network 23.1.1.0 0.0.0.255 area 0
R3(config-router)#network 3.3.3.3 0.0.0.0 area 3
// 特别注意，这个宣告时候设置的是 Area 3
R3(config-router)#exit
```

3. 查看 OSPF 的路由表，在 R1 上，具体如下所示。

```
R1#show ip route ospf
// 查看路由表中与 ospf 有关的路由信息
        2.0.0.0/32 is subnetted, 1 subnets
O IA    2.2.2.2 [110/2] via 12.1.1.2, 00:00:18, FastEthernet0/0
// "O IA" 表示的是 ospf 区域间路由，即学到的其他区域的路由
        3.0.0.0/32 is subnetted, 1 subnets
O IA    3.3.3.3 [110/2] via 13.1.1.3, 00:00:18, FastEthernet1/0
        23.0.0.0/24 is subnetted, 1 subnets
O       23.1.1.0 [110/2] via 13.1.1.3, 00:00:18, FastEthernet1/0
                 [110/2] via 12.1.1.2, 00:00:18, FastEthernet0/0
```

此时可以看到，R1 从 R2 和 R3 上学到其环回网段，并且路由类型发生变化，从"O"变成"O IA"，前者表示区域内路由，后者表示区域间路由。由于 R1、R2、R3 处于区域边界，所以是 ABR，可以通过命令查看，如下所示。

```
R1#show ip protocols
// 查看当前正在运行的路由协议的详细信息
Routing Protocol is "ospf 100"
    Outgoing update filter list for all interfaces is not set
    Incoming update filter list for all interfaces is not set
    Router ID 1.1.1.1
// 该路由器的 router-id 是 1.1.1.1
```

It is an area border router

// ABR:AREA BORDER ROUTER.连接多个区域的路由器是 ABR

Number of areas in this router is 2. 2 normal 0 stub 0 nssa

Maximum path: 4

Routing for Networks:

    1.1.1.1 0.0.0.0 area 1

    12.1.1.0 0.0.0.255 area 0

    13.1.1.0 0.0.0.255 area 0

Reference bandwidth unit is 100 mbps

Routing Information Sources:

Gateway	Distance	Last Update
3.3.3.3	110	00:06:46
2.2.2.2	110	00:06:46

Distance: (default is 110)

【ABR 路由器】不同区域之间的路由器，但是必须有一个区域是骨干区域。ABR 路由器的职责是用来实现不同区域间通信。区域间的 O IA 路由条目便是 ABR 通告的。

    4. 测试全网连通性，如下所示。

R1#ping 2.2.2.2 source 1.1.1.1

// 以 1.1.1.1 为源地址，去 PING 2.2.2.2

Type escape sequence to abort.

Sending 5, 100-byte ICMP Echos to 2.2.2.2, timeout is 2 seconds:

Packet sent with a source address of 1.1.1.1

!!!!!

// 成功

Success rate is 100 percent (5/5), round-trip min/avg/max = 20/27/36 ms

R1#ping 3.3.3.3 source 1.1.1.1

Type escape sequence to abort.

Sending 5, 100-byte ICMP Echos to 3.3.3.3, timeout is 2 seconds:

Packet sent with a source address of 1.1.1.1

!!!!!

// 成功

Success rate is 100 percent (5/5), round-trip min/avg/max = 16/22/28 ms

    同样的方法在 R2 和 R3 上也可以测试连通，说明通过部署 OSPF 多区域可以实现全网连通。此实验完成。

4.11 OSPF 路由汇总

实验目的：

掌握 OSPF 的区域汇总。

实验拓扑：

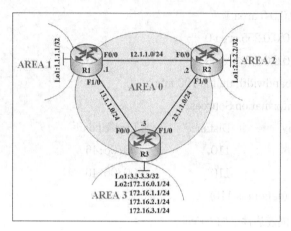

实验随手记：

_____
_____
_____
_____
_____

实验原理：

OSPF 路由汇总的出发点跟 RIP 和 EIGRP 一样，用于减小路由表体积并提高设备转发性能。但是 OSPF 的路由汇总是有条件的，具体如下。

①区域内路由器的链路状态数据库（LSDB）必须一致，所以同一区域内无法进行路由汇总。若执行汇总，则会造成区域内 LSDB 不对等。

②路由汇总只能在边界路由器（ABR 或者 ASBR）上执行。

③路由汇总是从常规区域往骨干区域的方向进行汇总。

实验步骤：

1. 依据图中拓扑配置各设备的 IP 地址，并保证直连连通性。

在 R1 上配置 IP 地址，如下所示。

4.11 OSPF 路由汇总

```
R1(config)#int f0/0
// 进入接口模式
R1(config-if)#no shutdown
// 打开接口
R1(config-if)#ip address 12.1.1.1 255.255.255.0
// 配置 IP 地址
R1(config-if)#exit
// 退出接口模式
R1(config)#int f1/0
R1(config-if)#no shutdown
R1(config-if)#ip address 13.1.1.1 255.255.255.0
R1(config-if)#exit
// 原理同上
R1(config)#int loopback 1
// 创建环回接口
R1(config-if)#ip address 1.1.1.1 255.255.255.255
// 配置环回 IP 地址
R1(config-if)#exit
```

在 R2 上配置 IP 地址，如下所示。

```
R2(config)#int f0/0
R2(config-if)#no shutdown
R2(config-if)#ip address 12.1.1.2 255.255.255.0
R2(config-if)#exit
R2(config)#int f1/0
R2(config-if)#no shutdown
R2(config-if)#ip address 23.1.1.2 255.255.255.0
R2(config-if)#exit
R2(config)#int loopback 1
R2(config-if)#ip address 2.2.2.2 255.255.255.255
R2(config-if)#exit
```

在 R3 上配置 IP 地址，如下所示。

```
R3(config)#int f0/0
R3(config-if)#no shutdown
R3(config-if)#ip address 13.1.1.3 255.255.255.0
R3(config-if)#exit
R3(config-if)#int f1/0
R3(config-if)#no shutdown
```

```
R3(config-if)#ip address 23.1.1.3 255.255.255.0
R3(config-if)#exit
R3(config)#int loopback 1
R3(config-if)#ip address 3.3.3.3 255.255.255.255
R3(config)#int loopback 2
R3(config-if)#ip add 172.16.0.1 255.255.255.0
R3(config-if)#ip add 172.16.1.1 255.255.255.0 secondary
R3(config-if)#ip add 172.16.2.1 255.255.255.0 secondary
R3(config-if)#ip add 172.16.3.1 255.255.255.0 secondary
R3(config-if)#exit
```

在其中一台路由器上进行连通性测试，如下所示。

```
R1#ping 12.1.1.2
// 在 R1 上面 PING R2 的 f0/0 端口来测试两个路由器之间的连通性
Type escape sequence to abort.
Sending 5, 100-byte ICMP Echos to 12.1.1.2, timeout is 2 seconds:
.!!!!
Success rate is 80 percent (4/5), round-trip min/avg/max = 28/36/48 ms
R1#ping 13.1.1.3
Type escape sequence to abort.
Sending 5, 100-byte ICMP Echos to 13.1.1.3, timeout is 2 seconds:
.!!!!
Success rate is 80 percent (4/5), round-trip min/avg/max = 16/31/48 ms
```

此时说明直连连接没有问题。

2. 在每台路由器上开始进行 OSPF 的配置，R1 的配置如下所示。

```
R1(config)#router ospf 100
// 在配置模式下开启 OSPF 路由协议，其中进程号是 100
R1(config-router)#router-id 1.1.1.1
// 设置 router-id 为 1.1.1.1，即用 1.1.1.1 这个地址来标识本路由器
R1(config-router)#network 12.1.1.0 0.0.0.255 area 0
// 用 Network 加 12.1.1.0 和反掩码（子网掩码取反）来宣告该条路由信息
R1(config-router)#network 13.1.1.0 0.0.0.255 area 0
R1(config-router)#network 1.1.1.1 0.0.0.0 area 1
// 特别注意，这个宣告时候设置的是 Area 1
R1(config-router)#exit
```

R2 的配置如下所示。

```
R2(config)#router ospf 100
R2(config-router)#router-id 2.2.2.2
R2(config-router)#network 12.1.1.0 0.0.0.255 area 0
R2(config-router)#network 23.1.1.0 0.0.0.255 area 0
R2(config-router)#network 2.2.2.2 0.0.0.0 area 2
// 特别注意宣告的是 Area 2
R2(config-router)#exit
```

R3 的配置如下所示。

```
R3(config)#router ospf 100
R3(config-router)#router-id 3.3.3.3
R3(config-router)#network 13.1.1.0 0.0.0.255 area 0
R3(config-router)#network 23.1.1.0 0.0.0.255 area 0
R3(config-router)#network 3.3.3.3 0.0.0.0 area 3
R3(config-router)#network 172.16.0.0 0.0.0.255 area 3
// 宣告路由信息，宣告在 Area 3
R3(config-router)#network 172.16.1.0 0.0.0.255 area 3
R3(config-router)#network 172.16.2.0 0.0.0.255 area 3
R3(config-router)#network 172.16.3.0 0.0.0.255 area 3
R3(config-router)#exit
```

3. 查看 OSPF 的路由表，在 R1 上，具体如下所示。

```
R1#show ip route ospf
     2.0.0.0/32 is subnetted, 1 subnets
O IA    2.2.2.2 [110/2] via 12.1.1.2, 00:01:56, FastEthernet0/0
     3.0.0.0/32 is subnetted, 1 subnets
O IA    3.3.3.3 [110/2] via 13.1.1.3, 00:01:56, FastEthernet1/0
23.0.0.0/24 is subnetted, 1 subnets
O       23.1.1.0 [110/2] via 13.1.1.3, 00:01:56, FastEthernet1/0
                 [110/2] via 12.1.1.2, 00:01:56, FastEthernet0/0
     172.16.0.0/16 is variably subnetted, 4 subnets, 2 masks
O IA    172.16.1.0/24 [110/2] via 13.1.1.3, 00:00:01, FastEthernet1/0
O IA    172.16.0.1/32 [110/2] via 13.1.1.3, 00:00:01, FastEthernet1/0
// 默认情况下，OSPF 会将环回接口当成末节主机，所以不管环回接口掩码是多少，都以 /32 掩码通告
O IA    172.16.2.0/24 [110/2] via 13.1.1.3, 00:00:01, FastEthernet1/0
O IA    172.16.3.0/24 [110/2] via 13.1.1.3, 00:00:01, FastEthernet1/0
```

在 R2 上，具体如下所示。

```
R2#show ip route ospf
// 查看路由表中有关 OSPF 协议的路由信息
        1.0.0.0/32 is subnetted, 1 subnets
O IA    1.1.1.1 [110/2] via 12.1.1.1, 00:34:34, FastEthernet0/0
        3.0.0.0/32 is subnetted, 1 subnets
O IA    3.3.3.3 [110/2] via 23.1.1.3, 00:34:34, FastEthernet1/0
        172.16.0.0/16 is variably subnetted, 4 subnets, 2 masks
O IA    172.16.1.0/24 [110/2] via 23.1.1.3, 00:00:57, FastEthernet1/0
O IA    172.16.0.1/32 [110/2] via 23.1.1.3, 00:00:57, FastEthernet1/0
O IA    172.16.2.0/24 [110/2] via 23.1.1.3, 00:00:57, FastEthernet1/0
O IA    172.16.3.0/24 [110/2] via 23.1.1.3, 00:00:57, FastEthernet1/0
```

此时 R1 和 R2 从 R3 学到区域间的 4 条精细路由。为了减少骨干区域的链路状态数据库，减小路由表体积并降低路由器的压力，可以采用区域间路由汇总技术。

4. 在 R3 上实现区域间路由汇总，如下所示。

```
R3(config)#router ospf 100
R3(config-router)# area 3 range 172.16.0.0 255.255.252.0
// Area + 精细路由所在区域 + 汇总路由
R3(config-router)#exit
//OSPF 的路由汇总有以下两个原则。
```

1. 不能在区域内汇总，区域内部路由器没法执行汇总功能，即便敲上命令也没有效果，必须是 ABR 或者 ASBR；
2. 一般汇总的方式都是从常规区域汇总到骨干区域。

再次查看 R1 和 R2 的路由表，具体如下所示。

```
R1#show ip route ospf
        2.0.0.0/32 is subnetted, 1 subnets
O IA    2.2.2.2 [110/2] via 12.1.1.2, 00:08:19, FastEthernet0/0
        3.0.0.0/32 is subnetted, 1 subnets
O IA    3.3.3.3 [110/2] via 13.1.1.3, 00:08:19, FastEthernet1/0
        23.0.0.0/24 is subnetted, 1 subnets
O       23.1.1.0 [110/2] via 13.1.1.3, 00:08:19, FastEthernet1/0
                 [110/2] via 12.1.1.2, 00:08:19, FastEthernet0/0
        172.16.0.0/22 is subnetted, 1 subnets
// 注意，这个时候已经汇总成功了，此时子网掩码已经变成 /22
O IA    172.16.0.0 [110/2] via 13.1.1.3, 00:00:47, FastEthernet1/0

R2#show ip route ospf
```

```
          1.0.0.0/32 is subnetted, 1 subnets
O IA      1.1.1.1 [110/2] via 12.1.1.1, 00:40:16, FastEthernet0/0
          3.0.0.0/32 is subnetted, 1 subnets
O IA      3.3.3.3 [110/2] via 23.1.1.3, 00:40:16, FastEthernet1/0
          172.16.0.0/22 is subnetted, 1 subnets
// 同样的，也在 R2 上汇总成功了
O IA      172.16.0.0 [110/2] via 23.1.1.3, 00:01:02, FastEthernet1/0
          13.0.0.0/24 is subnetted, 1 subnets
O         13.1.1.0 [110/2] via 23.1.1.3, 00:40:16, FastEthernet1/0
                   [110/2] via 12.1.1.1, 00:40:16, FastEthernet0/0
```

可以看到原本 4 条区域间精细路由汇总成 1 条，说明 R3 上区域间汇总成功。此实验完成。

第 5 章　交换技术

本章主要学习局域网交换技术，包括 VLAN、Trunk、VTP、STP、DHCP、Etherchannel、Port-Security 等技术。通过本章的学习，我们将掌握中小型局域网络的搭建。以下为本章导航图。

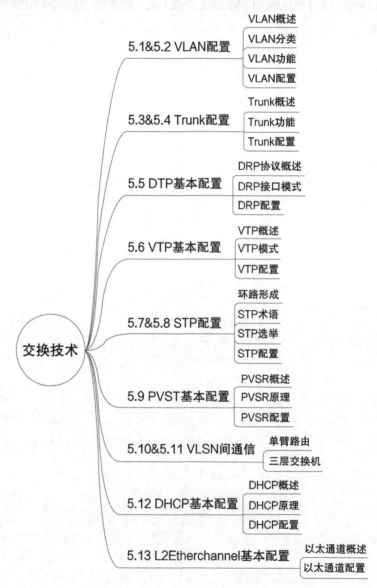

5.1 VLAN 基本配置

实验目的：
1. 掌握 VLAN 的基本配置。
2. 理解 VLAN 的功能。

实验拓扑：

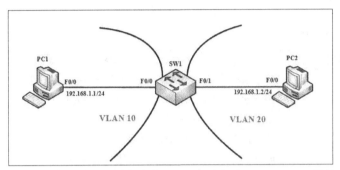

注明：本书所有交换实验通过在路由器 C3640 上加入 NM-16ESW 模块来进行模拟的，使用方法可以查看 GNS3 教程。若干交换实验需要通过真机或者 Cisco Packet Tracer 模拟器来实现。

实验随手记：

_____

_____

_____

_____

_____

实验原理：

1. VLAN 概述

VLAN(Virtual Local Area Network，虚拟局域网)技术可以将不同接口、主机、应用划分到不同网段，并且不同的 VLAN 处在不同的广播域，因此 VLAN 可以实现广播域隔离，减少广播流量泛洪，并且可以更好地对局域网设备进行管理。VLAN 具备跨交换机、跨地理位置的特性，同一 VLAN 的设备可以处在不同的交换机上，如图 5-1 所示。

【VLAN】
从本质上来讲，VLAN 便是将主机或服务器分组分域，便于管理。例如可以将公司的不同部门放入不同的 VLAN，后续可以针对不同的 VLAN 做不同的策略。默认情况下，不同 VLAN 之间的数据隔离，可以通过三层设备来进行 VLAN 间通信。

【广播域】
同一网段的主机处在相同的广播域之中，在同一广播域的环境下，只要有一台主机发送广播包，则整个广播域的其他主机都可以收到广播包。

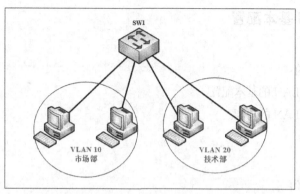

图 5-1　VLAN 原理

2. VLAN 分类

类型	描述
数据 VLAN	用于承载数据的 VLAN，是最常用的 VLAN 类型
语音 VLAN	用于承载语音流量，此 VLAN 下为语音设备
管理 VLAN	用于实现对交换机的管理
自然 VLAN	此 VLAN 的流量不需要打标签
私有 VLAN	此种 VLAN 可以实现 VLAN 内部安全隔离

3. VLAN 范围

Cisco 交换机的 VLAN 范围从 0 到 4095，不同范围的 VLAN 有不同的用途，如下表所示。

范围	描述
0，4095	系统保留 VLAN，不可使用和查看
1	系统默认 VLAN，所有接口默认在 VLAN1，不能修改和删除
2 ~ 1001	以太网下最常用的 VLAN 范围
1002 ~ 1005	用于 FDDI 和令牌环，不能修改和删除
1006 ~ 1024	系统保留 VLAN，不可使用和查看
1025 ~ 4094	拓展的以太网 VLAN

实验步骤：

1. 依据图中拓扑配置两台主机的 IP 地址，PC1 和 PC2 通过路由器模拟，配置如下所示。

```
PC1(config)#int f0/0
// 进入接口模式
PC1(config-if)#no shutdown
// 打开接口
PC1(config-if)#ip address 192.168.1.1 255.255.255.0
// 配置 IP 地址
PC1(config-if)#exit
```

```
PC2(config)#int f0/0
PC2(config-if)#no shutdown
PC2(config-if)#ip address 192.168.1.2 255.255.255.0
PC2(config-if)#exit
```

默认情况下，交换机的所有接口处于 VLAN1，如下所示。

```
SW1#show vlan-switch brief
// 查看 VLAN 的概要信息，在真实的交换机上可以通过 show vlan brief 直接查看
VLAN 信息
VLAN Name                        Status    Ports
---- -------------------------------- --------- -----------
1    default                     active    Fa0/0, Fa0/1, Fa0/2, Fa0/3
                                           Fa0/4, Fa0/5, Fa0/6, Fa0/7
                                           Fa0/8, Fa0/9, Fa0/10, Fa0/11
                                           Fa0/12, Fa0/13, Fa0/14, Fa0/15
// 可以看到，默认情况下，这些端口都是被置于 VLAN 1 的
1002 fddi-default                active
1003 token-ring-default          active
1004 fddinet-default             active
1005 trnet-default               active
```

此时在 PC1 上测试能否 Ping 通 PC2，如下所示。

```
PC1#ping 192.168.1.2
// 连接性测试，用 PC1 PING PC2
Type escape sequence to abort.
Sending 5, 100-byte ICMP Echos to 192.168.1.2, timeout is 2 seconds:
.!!!!
// 第一次失败，后面四次成功，则两个 PC 连通
Success rate is 100 percent (5/5), round-trip min/avg/max = 20/24/40 ms
```

从这里可以看出，由于 PC1 和 PC2 处于同一个 VLAN，所以能够相互通信。

2. 在交换机 SW1 上创建 VLAN10 和 VLAN20，并将不同的接口放入相应的 VLAN，如下所示。

```
SW1#vlan database
// 一般 3640 或者 3725 等系列路由器的交换模块需要进入 VLAN 数据库模式进行操作，
而一般的交换机则可以进入全局模式下部署。
(Config)#vlan 10
(config-vlan)#name VLAN_10
SW1(vlan)#vlan 10 name VLAN_10
// 创建 VLAN 并定义 VLAN 名字，一般不同的应用或者部门采用不同的命名，这样
易于管理
```

```
SW1(vlan)#vlan 20 name VLAN_20
SW1(vlan)#exit
APPLY completed.
Exiting....
SW1(config)#int f0/0
SW1(config-if)#switchport mode access
// 将接口模式修改为接入模式，此模式一般用于接入终端主机
SW1(config-if)#switchport access vlan 10
// 将接口放入 VLAN 10
SW1(config-if)#int f0/1
SW1(config-if)#switchport mode access
SW1(config-if)#switchport access vlan 20
SW1(config-if)#exit
```

查看 VLAN 信息，如下所示。

```
SW1#show vlan-switch brief

VLAN Name              Status    Ports
---- -------------------------------- --------- -----------
1    default           active   Fa0/2, Fa0/3, Fa0/4, Fa0/5
Fa0/6, Fa0/7, Fa0/8, Fa0/9
                      Fa0/10, Fa0/11, Fa0/12, Fa0/13
                      Fa0/14, Fa0/15
10   VLAN_10           active   Fa0/0
```
// 这个表示，Fa0/0 这个端口已经被成功放入 VLAN 10 了
```
20   VLAN_20           active   Fa0/1
1002 fddi-default               active
1003 token-ring-default         active
1004 fddinet-default            active
1005 trnet-default              active
```

此时，交换机生成 VLAN10 和 VLAN20，并且不同的接口处在不同的 VLAN 中。

3. 测试现在的 PC1 能否再次 Ping 通 PC2，如下所示。

```
PC1#ping 192.168.1.2

Type escape sequence to abort.
Sending 5, 100-byte ICMP Echos to 192.168.1.2, timeout is 2 seconds:
.....
```

```
// 可以看到，两者是不连通的，为什么呢？
Success rate is 0 percent (0/5)
```

从这里可以看出，由于不同的 VLAN 处于不同的广播域，并相互隔离，所以在默认情况下，不同 VLAN 间的设备无法通信。在项目环境下，一般不同的应用或企业的不同部门处于不同的 VLAN 中。若需要实现应用或者部门之间通信，则需要借助后续所学的单臂路由或三层交换来实现。此实验完成。

5.2 VLAN 进阶配置

实验目的：
1. 掌握跨交换机同 VLAN 的通信实现。
2. 理解跨交换机同 VLAN 的通信原理。

实验拓扑：

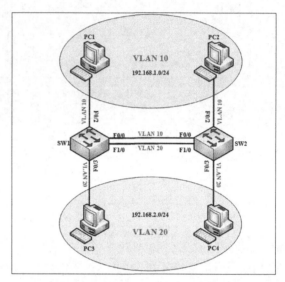

实验随手记：

_____

_____

_____

_____

_____

_____

实验步骤：

1. 依据图中拓扑配置 4 台主机的 IP 地址，其中 PC 通过路由器模拟，配置如下所示。

```
PC1(config)#int f0/0
// 进入接口模式
PC1(config-if)#no shutdown
// 打开接口
PC1(config-if)#ip address 192.168.1.1 255.255.255.0
```

```
// 配置 IP 地址
PC1(config-if)#exit
// 退出接口模式
PC2(config)#int f0/0
PC2(config-if)#no shutdown
PC2(config-if)#ip address 192.168.1.2 255.255.255.0
PC2(config-if)#exit
PC3(config)#int f0/0
PC3(config-if)#no shutdown
PC3(config-if)#ip address 192.168.2.3 255.255.255.0
PC3(config-if)#exit
PC4(config)#int f0/0
PC4(config-if)#no shutdown
PC4(config-if)#ip address 192.168.2.4 255.255.255.0
PC4(config-if)#exit
// 把 4 台 PC 的地址分别配置好
```

2. 根据图中拓扑，在交换机 SW1 和 SW2 上创建 VLAN，然后将接口放置到对应的 VLAN 中。

SW1 上配置如下所示。

```
SW1#vlan database
// 一般 3640 或者 3725 等系列路由器的交换模块需要进入 VLAN 数据库模式进行操作，而交换机则一般进入全局模式下部署。
(Config)#vlan 10
(config-vlan)#name VLAN_10
SW1(vlan)#vlan 10 name VLAN_10
// 创建 VLAN 并定义 VLAN 名字，一般不同的应用或者部门采用不同的命名，这样易于管理
SW1(vlan)#vlan 20 name VLAN_20
SW1(vlan)#exit
SW1(config)#int range f0/0 , f0/2
// 在接口前加 range 参数可以同时为多个接口做相同的配置
SW1(config-if-range)#switchport mode access
SW1(config-if-range)#switchport access vlan 10
SW1(config-if-range)#exit
SW1(config)#int range f0/1 , f0/3
SW1(config-if-range)#switchport mode access
SW1(config-if-range)#switchport access vlan 20
SW1(config-if-range)#exit
```

> // 在接口前加 range 参数可以同时为多个接口做相同的配置，案例如下。
> int range f0/0 - 2
> 表示从 f0/0 到 f0/2
> int range f0/0，f0/2
> 表示 f0/0 和 f0/2
> int range f0/0，f0/2 - 5
> 表示 f0/0 和 f0/2 到 f0/5

SW2 上配置如下所示。

```
SW2#vlan database
SW2(vlan)#vlan 10 name VLAN_10
SW2(vlan)#vlan 20 name VLAN_20
SW2(vlan)#exit
SW2(config)#int range f0/0，f0/2
SW2(config-if-range)#switchport mode access
SW2(config-if-range)#switchport access vlan 10
SW2(config-if-range)#exit
SW2(config)#int range f0/1，f0/3
SW2(config-if-range)#switchport mode access
SW2(config-if-range)#switchport access vlan 20
SW2(config-if-range)#exit
```

查看 VLAN 信息，如下所示。

```
SW1#show vlan-switch brief
// 查看 VLAN 的配置概要
VLAN Name              Status    Ports
---------------------------------------- --------------------
1    default           active    Fa0/4, Fa0/5, Fa0/6, Fa0/7
                                 Fa0/8, Fa0/9, Fa0/10, Fa0/11
                                 Fa0/12, Fa0/13,
Fa0/14, Fa0/15
10   VLAN_10           active    Fa0/0, Fa0/2
//Fa0/0 和 Fa0/2 都已经被放入 VLAN 10 中了
20   VLAN_20           active    Fa0/1, Fa0/3
1002 fddi-default                active
1003 token-ring-default          active
1004 fddinet-default             active
1005 trnet-default               active
```

```
SW2#show vlan-switch brief

VLAN Name              Status    Ports
---------------------------------------------------------
1    default           active    Fa0/4, Fa0/5, Fa0/6, Fa0/7
                                 Fa0/8, Fa0/9, Fa0/10, Fa0/11
                                 Fa0/12, Fa0/13, Fa0/14, Fa0/15
10   VLAN0010          active    Fa0/0, Fa0/2
20   VLAN0020          active    Fa0/1, Fa0/3
1002 fddi-default      active
1003 token-ring-default active
1004 fddinet-default   active
1005 trnet-default     active
```

此时，在 SW1 和 SW2 上已经创建了不同的交换机，并且不同的接口放置在对应的 VLAN 中。

3. 进行跨交换机同 VLAN 间的连通性测试，如下所示。

```
PC1#ping 192.168.1.2

Type escape sequence to abort.
Sending 5, 100-byte ICMP Echos to 192.168.1.1, timeout is 2 seconds:
.!!!!
// 通了！
Success rate is 80 percent (4/5), round-trip min/avg/max = 36/52/64 ms
PC3#ping 192.168.2.4
Type escape sequence to abort.
Sending 5, 100-byte ICMP Echos to 192.168.2.4, timeout is 2 seconds:
.!!!!
// 通了！
Success rate is 80 percent (4/5), round-trip min/avg/max = 32/43/60 ms
```

此时可以看出，同 VLAN 跨交换机通信成功！这个实验可以验证交换机之间承载多个 VLAN 流量可以通过多根网线划分到对应的 VLAN，从而实现通信。在 VLAN 数目比较少的情况下，这是一种简洁的解决方案。但是若 VLAN 数目比较多，则交换机之间需要连接的网线相应增多，此时则会消耗较多的交换机的端口资源，解决这个问题则需要引入接下来的 Trunk 技术。此实验完成。

5.3 Trunk 基本配置

实验目的：

1. 掌握 Trunk 的基础配置。
2. 理解 Trunk 的功能。

实验拓扑：

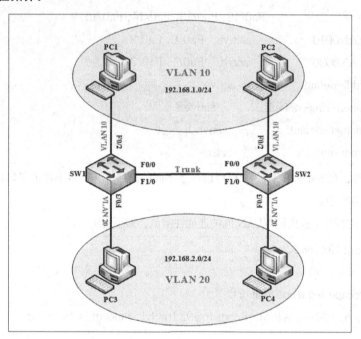

实验随手记：

_____
_____
_____
_____
_____
_____

实验原理：

1. Trunk 概述

当交换机之间存在多个 VLAN 的数据时，通过 Trunk 技术可以对不同 VLAN 的数据打上不同的标签，从而能很好地区分不同 VLAN 的数据，如图 5－2 所示。默认情况下，Trunk 技术可以承载所有 VLAN 的流量。

【Trunk】
Trunk 也被称为中继或者主干道，默认情况下可以同时承载局域网中所有 VLAN 的流量，Trunk 技术通过标签区分不同 VLAN 的数据流。Trunk 技术从逻辑上区分了 VLAN 数据流，节省了物理链路和端口资源。

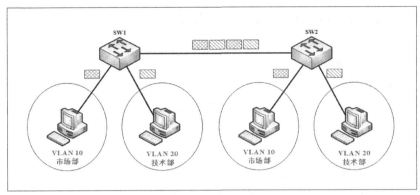

图 5-2 Trunk 原理

2. Trunk 封装

Trunk 采用两种封装协议，一种为 Cisco 私有的 ISL 协议（见图 5-3），一种为业界标准的 802.1Q（见图 5-4）。目前主流的封装方式为 802.1Q 封装。

| ISL TAG | DATA |

图 5-3 ISL 封装

图 5-4 802.1Q 封装

【Trunk 封装】Trunk 封装研究的是如何为一个数据包打上标签，默认最主流的方式是采用 802.1Q 的嵌入式封装。

两种封装协议的相关特性描述如下所示。

封装协议	描述
ISL	思科私有，采用包装式封装，标签字节为 26Bytes
802.1Q	业界标准，采用嵌入式封装，标签字节为 4Bytes

实验步骤：

1. 依据图中拓扑配置 4 台主机的 IP 地址，其中 PC 通过路由器模拟，配置如下所示。

PC1(config)#int f0/0

// 进入接口模式

PC1(config-if)#no shutdown

// 打开接口

PC1(config-if)#ip address 192.168.1.1 255.255.255.0

// 配置 IP 地址

PC1(config-if)#exit

// 退出接口模式

PC2(config)#int f0/0

PC2(config-if)#no shutdown

```
PC2(config-if)#ip address 192.168.1.2 255.255.255.0
PC2(config-if)#exit
PC3(config)#int f0/0
PC3(config-if)#no shutdown
PC3(config-if)#ip address 192.168.2.3 255.255.255.0
PC3(config-if)#exit
PC4(config)#int f0/0
PC4(config-if)#no shutdown
PC4(config-if)#ip address 192.168.2.4 255.255.255.0
PC4(config-if)#exit
```

2. 根据图中拓扑，在交换机 SW1 和 SW2 上创建 VLAN，然后将接口放置到对应的 VLAN 中。

SW1 上配置如下所示。

```
SW1#vlan database
// 一般 3640 或者 3725 等系列路由器的交换模块需要进入 VLAN 数据库模式中进行操作
SW1(vlan)#vlan 10 name VLAN_10
// 创建 VLAN 并定义 VLAN 名字，一般不同的应用或者部门采用不同的命名，这样易于管理
SW1(vlan)#vlan 20 name VLAN_20
SW1(vlan)#exit
SW1(config)#int f0/2
SW1(config-if)#switchport mode access
// 将接口模式修改为接入模式，此模式一般用于接入终端主机
SW1(config-if)#switchport access vlan 10
// 将接口放入 VLAN 10 中
SW1(config-if-range)#exit
SW1(config)#int f0/3
SW1(config-if-range)#switchport mode access
SW1(config-if-range)#switchport access vlan 20
SW1(config-if-range)#exit
```

SW2 上配置如下所示。

```
SW2#vlan database
SW2(vlan)#vlan 10 name VLAN_10
SW2(vlan)#vlan 20 name VLAN_20
SW2(vlan)#exit
SW2(config)#int f0/2
```

```
SW2(config-if-range)#switchport mode access
SW2(config-if-range)#switchport access vlan 10
SW2(config-if-range)#exit
SW2(config)#int f0/3
SW2(config-if-range)#switchport mode access
SW2(config-if-range)#switchport access vlan 20
SW2(config-if-range)#exit
```

查看 VLAN 信息，如下所示。

```
SW1#show vlan-switch brief
// 查看交换机上 VLAN 的概要信息
VLAN Name              Status    Ports
---- -------------------------------- --------- -----------
1    default             active Fa0/1,Fa0/4, Fa0/5, Fa0/6, Fa0/7
                                Fa0/8, Fa0/9, Fa0/10, Fa0/11
                                Fa0/12, Fa0/13, Fa0/14, Fa0/15
10   VLAN_10             active   Fa0/2
//Fa0/2 这个端口已经放入 VLAN_10 内
20   VLAN_20             active   Fa0/3
//Fa0/3 这个端口已经放入 VLAN_20 内
1002 fddi-default             active
1003 token-ring-default       active
1004 fddinet-default          active
1005 trnet-default            active
SW2#show vlan-switch brief

VLAN Name              Status    Ports
---- -------------------------------- --------- -----------
1    default             active Fa0/1,Fa0/4, Fa0/5, Fa0/6, Fa0/7
                                Fa0/8, Fa0/9, Fa0/10, Fa0/11
                                Fa0/12, Fa0/13, Fa0/14, Fa0/15
10   VLAN0010            active   Fa0/2
20   VLAN0020            active   Fa0/3
1002 fddi-default             active
1003 token-ring-default       active
1004 fddinet-default          active
1005 trnet-default            active
```

此时，在 SW1 和 SW2 上已经创建了不同的交换机，并且不同的接口放置在对应的 VLAN 中。

3. 部署 Trunk 技术，SW1 上配置如下所示。

SW1(config)#int f0/0
SW1(config-if)#switchport trunk encapsulation dot1q
// Trunk 有两种封装标准，一种是 Cisco 私有的 ISL，一种为行业标准 802.1Q，一般采用 802.1Q 实现封装
SW1(config-if)#switchport mode trunk
// 将接口模式定义为 trunk 模式，一般交换机相连的接口采用 trunk 模式，用于承载不同 VLAN 的流量。
SW1(config-if)#exit

SW2 上配置如下所示。

SW2(config)#int f0/0
SW2(config-if)#switchport trunk encapsulation dot1q
// Trunk 有两种封装标准，一种是 Cisco 私有的 ISL，一种为行业标准 802.1Q，一般采用 802.1Q 实现封装
SW2(config-if)#switchport mode trunk
// 将接口模式定义为 trunk 模式，一般交换机相连的接口采用 trunk 模式，用于承载不同 VLAN 的流量。
SW2(config-if)#exit

查看 Trunk 链路状态，如下所示。

SW1#show interfaces trunk
// 这条命令用于查看 Trunk 信息

Port   Mode   Encapsulation Status   Native vlan
Fa0/0   on   802.1q   trunking   1
// 一般看 "Trunk" 则表示 Trunk 链路起来了

Port   Vlans allowed on trunk
Fa0/0   1-1005

Port   Vlans allowed and active in management domain
Fa0/0   1,10,20

Port   Vlans in spanning tree forwarding state and not pruned
Fa0/0   1,10,20

此时 Trunk 链路已经起来。

4. 测试通过部署 Trunk 链路，VLAN10 和 VLAN20 的数据能否经过 Trunk，如下所示。

```
PC1#ping 192.168.1.2
// 用 PC1 PING PC2
Type escape sequence to abort.
Sending 5, 100-byte ICMP Echos to 192.168.1.2, timeout is 2 seconds:
!!!!!
// 通了
Success rate is 100 percent (5/5), round-trip min/avg/max = 32/42/64 ms
PC3#ping 192.168.2.4

Type escape sequence to abort.
Sending 5, 100-byte ICMP Echos to 192.168.2.4, timeout is 2 seconds:
!!!!!
// 通了
Success rate is 100 percent (5/5), round-trip min/avg/max = 44/56/68 ms
```

从以上可以看到，Trunk 链路可以同时承载不同 VLAN 的流量，相比之前 "一个 VLAN 一根网线" 的解决方案，Trunk 方案不仅部署方便，而且非常节省端口资源，做到 "n 个 VLAN 一根网线"。当然，由于 Trunk 链路默认承载所有 VLAN 的流量，因而会有很多垃圾流量在上面，这就需要实现 Trunk 的优化。此实验完成。

5.4 Trunk 进阶配置

实验目的:
1. 掌握 Native vlan 和 Allow vlan 的配置。
2. 理解 Native vlan 和 Allow vlan 的功能。

实验拓扑:

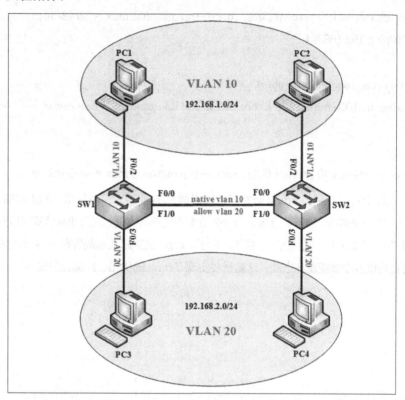

实验随手记:

_____
_____
_____
_____
_____
_____

实验原理:

1. Native vlan

在 Trunk 技术中,vlan 的数量和 tag(标签)的数量是一致的,不同的 vlan 对应不同的 tag。而 native vlan 是一种特殊的 vlan,从 native vlan 出来的数据包,在 trunk 链路上不做标签封装,整个 trunk 链路只允许一个 native vlan 存在,默认为

【Native VLAN】
自然 VLAN 也称为本征 VLAN,是 Trunk 封装的优化技术。

vlan1。一般情况下，将网络中大数据流所在的 vlan 设置为 native vlan，这样可以有效降低交换机打 tag 的资源消耗以提高转发效率。

2. Allowed vlan

Trunk 链路默认承载所有 vlan 的流量，但是可以通过 allowed vlan 参数来进行选择性承载，可以只允许某些 vlan 的流量通过 trunk，如图 5-5 所示。

【Allow VLAN】
允许 VLAN 也是 Trunk 的优化技术，用于有选择地进行 VLAN 的数据转发。

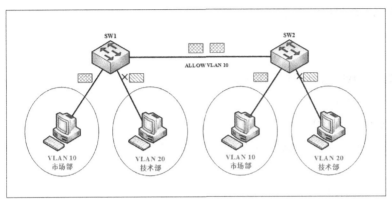

图 5-5　allow vlan 原理

实验步骤：

1. 依据图中拓扑配置 4 台主机的 IP 地址，其中 PC 通过路由器模拟，配置如下所示。

```
PC1(config)#int f0/0
// 进入接口模式
PC1(config-if)#no shutdown
// 打开接口
PC1(config-if)#ip address 192.168.1.1 255.255.255.0
// 配置 IP 地址
PC1(config-if)#exit
// 退出接口模式
PC2(config)#int f0/0
PC2(config-if)#no shutdown
PC2(config-if)#ip address 192.168.1.2 255.255.255.0
PC2(config-if)#exit
PC3(config)#int f0/0
PC3(config-if)#no shutdown
PC3(config-if)#ip address 192.168.2.3 255.255.255.0
PC3(config-if)#exit
PC4(config)#int f0/0
PC4(config-if)#no shutdown
PC4(config-if)#ip address 192.168.2.4 255.255.255.0
PC4(config-if)#exit
```

2. 根据图中拓扑，在交换机 SW1 和 SW2 上创建 VLAN，然后将接口放置到对应的 VLAN 中。

SW1 上配置如下所示。

```
SW1#vlan database
// 一般 3640 或者 3725 等系列路由器的交换模块需要进入 VLAN 数据库模式进行操作
SW1(vlan)#vlan 10 name VLAN_10
// 创建 VLAN 并定义 VLAN 名字，一般不同的应用或者部门采用不同的命名，这样易于管理
SW1(vlan)#vlan 20 name VLAN_20
SW1(vlan)#exit
SW1(config)#int f0/2
SW1(config-if)#switchport mode access
// 将接口模式修改为接入模式，此模式一般用于接入终端主机
SW1(config-if)#switchport access vlan 10
// 将接口放入 VLAN 10 中
SW1(config-if-range)#exit
SW1(config)#int f0/3
SW1(config-if-range)#switchport mode access
SW1(config-if-range)#switchport access vlan 20
SW1(config-if-range)#exit
```

SW2 上配置如下所示。

```
SW2#vlan database
SW2(vlan)#vlan 10 name VLAN_10
SW2(vlan)#vlan 20 name VLAN_20
SW2(vlan)#exit
SW2(config)#int f0/2
SW2(config-if-range)#switchport mode access
SW2(config-if-range)#switchport access vlan 10
SW2(config-if-range)#exit
SW2(config)#int f0/3
SW2(config-if-range)#switchport mode access
SW2(config-if-range)#switchport access vlan 20
SW2(config-if-range)#exit
```

查看 VLAN 信息，如下所示。

```
SW1#show vlan-switch brief
// 查看交换机上 VLAN 的概要信息
VLAN Name                    Status    Ports
```

```
---- ------------------------------ --------- ----------
1    default                        active    Fa0/1,Fa0/4, Fa0/5, Fa0/6, Fa0/7
                                              Fa0/8, Fa0/9, Fa0/10, Fa0/11
                                              Fa0/12, Fa0/13, Fa0/14, Fa0/15
10   VLAN_10                        active    Fa0/2
//Fa0/2 这个端口已经放入 VLAN_10 内
20   VLAN_20                        active    Fa0/3
//Fa0/3 这个端口已经放入 VLAN_20 内
1002 fddi-default                   active
1003 token-ring-default             active
1004 fddinet-default                active
1005 trnet-default                  active
SW2#show vlan-switch brief

VLAN Name                           Status    Ports
---- ------------------------------ --------- ----------
1    default                        active    Fa0/1,Fa0/4, Fa0/5, Fa0/6, Fa0/7
                                              Fa0/8, Fa0/9, Fa0/10, Fa0/11
                                              Fa0/12, Fa0/13, Fa0/14, Fa0/15
10   VLAN_10                        active    Fa0/2
20   VLAN_20                        active    Fa0/3
1002 fddi-default                   active
1003 token-ring-default             active
1004 fddinet-default                active
1005 trnet-default                  active
```

此时，SW1和SW2上不同交换机已经创建，并且不同接口放置在对应的VLAN中。

3. 部署 Trunk 技术，并实现 Trunk 的优化。默认情况下，Trunk 上 native vlan 为 1，即从 vlan1 的数据不打标签，要求将 native vlan 改为 10；其次，默认 Trunk 允许所有的 vlan 数据通过，要求只允许 vlan 10 和 20 通过。配置如下所示。

```
SW1(config)#int f0/0
SW1(config-if)#switchport trunk encapsulation dot1q
// Trunk 有两种封装标准，一种是 Cisco 私有的 ISL，一种为行业标准 802.1Q，一般
采用 802.1Q 实现封装
SW1(config-if)#switchport mode trunk
```

// 将接口模式定义为 trunk 模式，一般交换机相连的接口采用 trunk 模式，用于承载不同 VLAN 的流量

SW1(config-if)#switchport trunk native vlan 10

// 将默认的 Native vlan 从 1 修改为 10

SW1(config-if)#switchport trunk allowed 1,1002-1005,10,20

// 默认 trunk 允许所有 vlan，此处修改为只允许 10 和 20；除此之外还需将系统默认的 1,1002-1005 加入，一般交换机则可以部署以下命令实现。

Switchport trunk allowed 10,20

SW1(config-if)#exit

SW2 上配置如下所示。

SW2(config)#int f0/0
SW2(config-if)#switchport trunk encapsulation dot1q
SW2(config-if)#switchport mode trunk
SW2(config-if)#switchport trunk native vlan 10
SW2(config-if)#switchport trunk allowed vlan 1,1002-1005,10,20
SW2(config-if)#exit

查看 Trunk 链路状态，如下所示。

SW1#show interfaces trunk

// 查看接口 Trunk 信息

Port   Mode   Encapsulation   Status   Native vlan
Fa0/0   on   802.1q   trunking 10

//Native vlan 变成了 VLAN 10 了

Port   Vlans allowed on trunk
Fa0/0   1,10,20,1002-1005

// 此接口只允许指定的这些 VLAN 通过

Port   Vlans allowed and active in management domain
Fa0/0   1,10,20

Port   Vlans in spanning tree forwarding state and not pruned
Fa0/0   1,10,20

可以看到，native vlan 从 1 变成 10，而 allow vlan 则只允许 vlan10、20 和其他默认 vlan 数据通过。

4. 进入 Trunk 优化测试，要验证 native vlan 的效果，可以通过抓包来达到，例如先让 PC1 ping PC2，并在 trunk 上抓包。

PC1#ping 192.168.1.2

//PING 通

Type escape sequence to abort.

Sending 5, 100-byte ICMP Echos to 192.168.1.2, timeout is 2 seconds:

!!!!!

// 成功

Success rate is 100 percent (5/5), round-trip min/avg/max = 28/42/64 ms

底层数据进行分组，如图 5-6 所示。

图 5-6　Trunk 抓包

再让 PC3 ping PC4，并抓包，如下所示。

PC3#ping 192.168.2.4

Type escape sequence to abort.

Sending 5, 100-byte ICMP Echos to 192.168.2.4, timeout is 2 seconds:

!!!!!

Success rate is 100 percent (5/5), round-trip min/avg/max = 28/40/60 ms

底层数据进行分组，如图 5-7 所示。

图 5-7　Trunk 抓包

从上面对比可以看出，一般的 vlan 经过 trunk 链路需要打上标签，而 native vlan 无需打上标签。

5. 验证 Allow vlan 功能，将允许的 vlan 改为只允许 vlan20 通过，配置如下所示。

SW1(config)#int f0/0

SW1(config-if)#switchport trunk allowed vlan 1,20,1002-1005

SW1(config-if)#exit

SW2(config)#int f0/0

SW2(config-if)#switchport trunk allowed vlan 1,20,1002-1005

SW2(config-if)#exit

此时让 PC1 ping PC2，如下所示。

PC1#ping 192.168.1.2

Type escape sequence to abort.

Sending 5, 100-byte ICMP Echos to 192.168.1.2, timeout is 2 seconds:

```
......
// 不能通过了
Success rate is 0 percent (0/5)
```

再让 PC3 ping PC4，如下所示。

```
PC3#ping 192.168.2.4
Type escape sequence to abort.
Sending 5, 100-byte ICMP Echos to 192.168.2.4, timeout is 2 seconds:
!!!!!

Success rate is 100 percent (5/5), round-trip min/avg/max = 24/36/60 ms
```

从上面可以看出，没有被 allow 的 vlan 没法通过 trunk 链路！

通过以上两种 trunk 优化的部署，我们可以得出：

① Native vlan 可以使特定的 vlan 在经过 trunk 的时候无需打上标签，交换机只允许一个 native vlan，默认为 native vlan 1。一般将 native vlan 修改为需要大数据处理的 vlan，由此可以减轻交换机的压力。另外，交换机双方的 native vlan 必须一致，否则，由于 Cisco 交换机开启 CDP 协议，若检测到不一致，则链路会 down！

② Allow vlan 可以使特定的 vlan 在 trunk 上面跑，通过此技术，可以限制一些垃圾数据如广播泛洪的影响，以达到流量优化。此实验完成。

5.5 DTP 基本配置

实验目的：
1. 掌握 DTP 的基本配置。
2. 理解 DTP 的接口模式及原理。

实验拓扑：

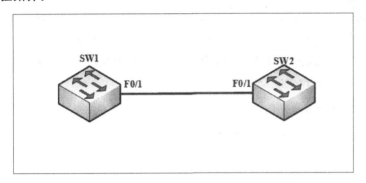

实验随手记：

_____
_____
_____
_____
_____
_____
_____

实验原理：

1. DTP 概述

DTP（Dynamic Trunking Protocol，动态中继协议）技术是思科私有的技术，用于动态协商 Trunk 链路。

2. DTP 原理

DTP 将接口设置为不同的协商模式，不同的组合将决定 Trunk 链路是否协商成功。在实验步骤中有表格说明。

实验步骤：

1. DTP 协商模式主要分为四种，具体如下。

DD<dynamic Desirable>：主动模式，接口发送 DTP 分组进行协商。

DA<dynamic Auto>：被动模式，接口接收 DTP 分组进行协商。

【DTP 协议】
由于 DTP 协议是思科私有协议，所以不同厂商之间无法同时采用 DTP 协议进行 Trunk 协商，此时可以关闭 DTP 协商并强制开启 Trunk。

ON：一般接口强制配置为 trunk，则为 on 模式。

OFF：一般接口强制配置为 access，则为 off 模式。

根据这几种模式，可以有不同的协商结果，如下表所示。

	Dynamic Auto	Dynamic Desirable	Trunk/on	Access/off
Dynamic Auto	Access	Trunk	Trunk	Access
Dynamic Desirable	Trunk	Trunk	Trunk	Access
Trunk/on	Trunk	Trunk	Trunk	Not recommended
Access/off	Access	Access	Not recommended	Access

2. 依据以上表在 SW1 和 SW2 上部署 DTP 协议，并观看 Trunk 链路的协商状态。

①范例一：DD & DD，如下所示。

```
SW1(config)#int f0/1
SW1(config-if)#switchport trunk encapsulation dot1q
// Trunk 有两种封装标准，一种是 Cisco 私有的 ISL，另一种是行业标准 802.1Q，一般采用 802.1Q 实现封装
SW1(config-if)#switchport mode dynamic desirable
// 将接口配置为动态协商主动模式
SW1(config-if)#exit
SW2(config)#int f0/1
SW2(config-if)#switchport trunk encapsulation dot1q
SW2(config-if)#switchport mode dynamic desirable
SW2(config-if)#exit
```

②范例二：DD & DA，如下所示。

```
SW1(config)#int f0/1
SW1(config-if)#switchport trunk encapsulation dot1q
SW1(config-if)#switchport mode dynamic desirable
SW1(config-if)#exit
SW2(config)#int f0/1
SW2(config-if)#switchport trunk encapsulation dot1q
SW2(config-if)#switchport mode dynamic auto
// 将接口配置为动态协商被动模式
SW2(config-if)#exit
```

③范例三：DA & DA，如下所示。

```
SW1(config)#int f0/1
SW1(config-if)#switchport trunk encapsulation dot1q
SW1(config-if)#switchport mode dynamic auto
SW1(config-if)#exit
```

```
SW2(config)#int f0/1
SW2(config-if)#switchport trunk encapsulation dot1q
SW2(config-if)#switchport mode dynamic auto
SW2(config-if)#exit
```

3. 工程用法。从以上可以看出，DTP 可以通过不同模式进行 Trunk 的协商，但是 DTP 属于 CISCO 的私有技术，若网络中有其他厂商的交换机则没法进行协商。例如，接入层交换机用的是 H3C 交换机，汇聚层用的是 CISCO 交换机，则 DTP 失去意义。所以，一般工程用法习惯直接将接口强制设置为 trunk，并关闭 DTP 协商，如下所示。

```
SW1(config)#int f0/1
SW1(config-if)#switchport trunk encapsulation dot1q
SW1(config-if)#switchport mode trunk
SW1(config-if)#switchport nonegotiate
// 关闭 DTP 协商
SW2(config)#int f0/1
SW2(config-if)#switchport trunk encapsulation dot1q
SW2(config-if)#switchport nonegotiate
```

此实验完成。

5.6 VTP 基本配置

实验目的：

1. 掌握 VTP 的基本配置。
2. 理解 VTP 的 Server、Client 模式。
3. 理解 VTP 的管理域、密码、配置版本号。

实验拓扑：

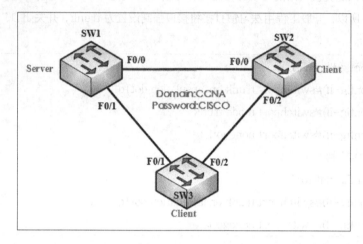

实验随手记：

_____
_____
_____
_____
_____
_____

【VTP 协议】
VTP 协议可以实现 VLAN 的统一管理，包括创建、修改、删除 VLAN。

【配置版本号】
配置版本号是用于衡量 VLAN 信息新旧程度的数值，数值越高权限越高。

实验原理：

1. VTP 概述

VTP（Vlan Trunking Protocol，vlan 中继协议）技术是思科私有的技术，用于实现对 VLAN 的全局管理（创建、删除、修改）。VTP 可以有效提高局域网 VLAN 的管理效率，但是由于是 Cisco 私有的技术，在多厂商环境下就没法发挥出它的功能。

2. VTP 原理

VTP 技术通过在交换机中设置 Server 和 Client 模式，由 Server 统一进行 VLAN 管理，Client 只能进行转发和接收。Server 和 Client 的域名和密码必须一致，否则无法进行 VLAN 同步。默认情况下，所有 Server 和 Client 的配置版本号为 0，当 VLAN 信息发生变动时（如创建 VLAN），则版本号增加 1。版本号越高权限越大，版本号低的设备需要向版本号高的设备进行同步。一般 Server 端的版本号比 Client

端的版本号高，所以 Client 向 Server 学习，如图 5-8 所示。

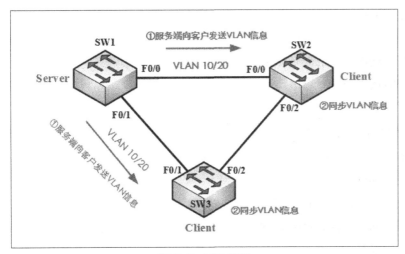

图 5-8　VTP 原理

3. VTP 模式

模式	描述
Server	能创建、删除、修改 VLAN
Client	不能创建、删除、修改 VLAN，能学习转发
Transparent	能创建、删除、修改 VLAN，不能学习转发

实验步骤：

1. 依据图中拓扑部署 VTP 技术，整个 VTP 管理域为 CCNA，密码为 Cisco，其中 SW1 为 Server，其他交换机为 Client。

SW1 上配置如下所示。

```
SW1#vlan database
//3640 或者 3725 等系列路由器的交换模块需要进入 VLAN 数据库模式进行操作

SW1(vlan)#vtp server
// 定义 VTP 模式，全局模式方式：
     (config)vtp mode server/client

SW1(vlan)#vtp domain CCNA
// 定义 VTP 管理域，与全局模式配置方法一致
SW1(vlan)#vtp password Cisco
// 定义 VTP 密码，实现 VTP 安全认证，与全局模式配置方法一致
// 域名和密码需要一致才能进行通信
SW1(vlan)#exit
```

SW2 上配置如下所示。

```
SW2#vlan database
SW2(vlan)#vtpclient
// 特别注意，这里是 Client 模式
SW2(vlan)#vtp domain CCNA
SW2(vlan)#vtp password Cisco
SW2(vlan)#exit
```

SW3 上配置如下所示。

```
SW3#vlan database
SW3(vlan)#vtp client
SW3(vlan)#vtp domain CCNA
SW3(vlan)#vtp password Cisco
SW3(vlan)#exit
```

此时查看 VTP 信息，如下所示。

```
SW1#show vtp status
// 查看 VTP 状态
VTP Version                    : 2
Configuration Revision         : 0
```

// 此处显示配置版本号，默认为 0，每次 VLAN 信息的变动都会增加 1，配置版本号越大越优先，配置版本号小的交换机会向配置版本号高的交换机自动同步 VLAN 信息

```
Maximum VLANs supported locally : 256
Number of existing VLANs       : 5
```

// 本地存在的 VLAN 数目，默认有系统 VLAN5 个，即
VLAN1 和 1002 到 1005，并且没法修改和删除。

```
VTP Operating Mode             : Server
```

//VTP 操作模式，有 Server、Client、Transparent 三种，默认所有交换机都是 Server 模式。

```
VTP Domain Name                : CCNA
```

//VTP 域名，大小写敏感

```
VTP Pruning Mode               : Disabled
VTP V2 Mode                    : Disabled
VTP Traps Generation           : Disabled
MD5 digest                     : 0x4E 0xC8 0xE0 0x2D 0x31 0x2E 0x45 0x01
Configuration last modified by 0.0.0.0 at 3-1-02 01:22:27
Local updater ID is 0.0.0.0 (no valid interface found)
```

```
SW2#show vtp status
VTP Version                       : 2
Configuration Revision            : 0
Maximum VLANs supported locally   : 256
Number of existing VLANs          : 5
VTP Operating Mode                : Client
VTP Domain Name                   : CCNA
VTP Pruning Mode                  : Disabled
VTP V2 Mode                       : Disabled
VTP Traps Generation              : Disabled
MD5 digest                        : 0x4E 0xC8 0xE0 0x2D 0x31 0x2E 0x45 0x01
Configuration last modified by 0.0.0.0 at 3-1-02 01:22:27
```

可以看出，由于此时 Server 端未创建 VLAN，默认配置版本号都为 0。

2. 为了使得全网都能学习到 VLAN10/20/30，需要在 Server 创建，配置如下所示。

```
SW1(vlan)#vlan 10 name VLAN_10
// 创建 VLAN 10 并命名为 VLAN_10
SW1(vlan)#vlan 20 name VLAN_20
SW1(vlan)#vlan 30 name VLAN_30
SW1(vlan)#exit
```

此时查看 SW1 上 VTP 信息，如下所示。

```
SW1#show vtp status
VTP Version                       : 2
Configuration Revision            : 1
// 版本增加 1 了
Maximum VLANs supported locally   : 256
Number of existing VLANs          : 8
// 多增加了 3 个 VLAN，是刚刚手工加入的
VTP Operating Mode                : Server
VTP Domain Name                   : CCNA
VTP Pruning Mode                  : Disabled
VTP V2 Mode                       : Disabled
VTP Traps Generation              : Disabled
MD5 digest                        : 0x46 0xD6 0x5A 0x38 0xF2 0xFD 0xA5 0xE6
Configuration last modified by 0.0.0.0 at 3-1-02 01:39:10
Local updater ID is 0.0.0.0 (no valid interface found)
```

由于 VLAN 信息出现变动，所以 VTP 版本号增加，此时查看其他 Client 端信息，如下所示。

```
SW2#show vtp status
VTP Version                       : 2
Configuration Revision            : 1
Maximum VLANs supported locally   : 256
Number of existing VLANs          : 8
VTP Operating Mode                : Client
VTP Domain Name                   : CCNA
VTP Pruning Mode                  : Disabled
VTP V2 Mode                       : Disabled
VTP Traps Generation              : Disabled
MD5 digest                        : 0x46 0xD6 0x5A 0x38 0xF2 0xFD 0xA5 0xE6
Configuration last modified by 0.0.0.0 at 3-1-02 01:39:10
```

可以看到，由于 Server 端版本号较高，Client 端已经从 Server 端同步 VLAN 信息，查看 VLAN 信息，如下所示。

```
SW2#show vlan-switch brief

VLAN Name                        Status    Ports
---- -------------------------------- --------- -------------
1    default                     active    Fa0/1, Fa0/3, Fa0/4, Fa0/5
                                           Fa0/6, Fa0/7, Fa0/8, Fa0/9
                                           Fa0/10, Fa0/11, Fa0/12, Fa0/13
                                           Fa0/14, Fa0/15
10   VLAN_10                     active
20   VLAN_20                     active
30   VLAN_30                     active
// 已经成功同步
1002 fddi-default                active
1003 token-ring-default          active
1004 fddinet-default             active
1005 trnet-default               active
```

同样在 SW3 上查看 VLAN 信息，如下所示。

```
SW3#show vlan-switch brief
```

```
VLAN Name                    Status    Ports
---- -------------------------------- --------- ----------
1    default                 active    Fa0/0, Fa0/3, Fa0/4, Fa0/5
                                       Fa0/6, Fa0/7, Fa0/8, Fa0/9
                                       Fa0/10, Fa0/11, Fa0/12, Fa0/13
                                       Fa0/14, Fa0/15
10   VLAN_10                 active
20   VLAN_20                 active
30   VLAN_30                 active
1002 fddi-default            active
1003 token-ring-default      active
1004 fddinet-default         active
1005 trnet-default           active
```

可以看到，通过部署 VTP 技术，此时 VLAN 信息全局同步。

3. 修改 SW2 和 SW3 的管理域或者密码，并在 SW1 上修改 VLAN 信息，查看 VTP 同步效果。

SW2 上修改域名，如下所示。

```
SW2(vlan)#vtp domain PL
// 修改域名
Changing VTP domain name from CCNA to PL
// 提示成功
SW2(vlan)#exit
```

SW3 上修改密码，如下所示。

```
SW3(vlan)#vtp password PL
// 修改密码
Setting device VLAN database password to PL.
SW3(vlan)#exit
```

SW1 上修改 VLAN，如下所示。

```
SW1#vlan database
SW1(vlan)#vlan 10 name TEST
SW1(vlan)#no vlan 20
SW1(vlan)#exit
```

此时查看 SW1 上 VLAN 信息，如下所示。

```
SW1#show vlan-switch brief

VLAN Name                    Status    Ports
---- ------------------------------ --------- -------------------------------
1    default                 active   Fa0/2, Fa0/3, Fa0/4, Fa0/5
                                      Fa0/6, Fa0/7, Fa0/8, Fa0/9
                                      Fa0/10, Fa0/11, Fa0/12, Fa0/13
                                      Fa0/14, Fa0/15
10   TEST                    active
1002 fddi-default            active
1003 token-ring-default      active
1004 fddinet-default         active
1005 trnet-default           active
```

再次观察 SW2 和 SW3 上的 VLAN 信息，如下所示。

```
SW2#show vlan-switch brief

VLAN Name                    Status    Ports
---- ------------------------------ --------- -------------------------------
1    default                 active   Fa0/1, Fa0/3, Fa0/4, Fa0/5
                                      Fa0/6, Fa0/7, Fa0/8, Fa0/9
                                      Fa0/10, Fa0/11, Fa0/12, Fa0/13
                                      Fa0/14, Fa0/15
10   VLAN_10                 active
20   VLAN_20                 active
30   VLAN_30                 active

// 因为域名和密码都不对，所以就不再交互 VTP 信息了
1002 fddi-default            active
1003 token-ring-default      active
1004 fddinet-default         active
1005 trnet-default           active

SW3#show vlan-switch brief

VLAN Name                    Status    Ports
```

1 default	active	Fa0/0, Fa0/3, Fa0/4, Fa0/5
		Fa0/6, Fa0/7, Fa0/8, Fa0/9
		Fa0/10, Fa0/11, Fa0/12, Fa0/13
		Fa0/14, Fa0/15
10 VLAN_10	active	
20 VLAN_20	active	
30 VLAN_30	active	

// 因为域名和密码都不对，所以就不再交互 VTP 信息了

1002 fddi-default	active
1003 token-ring-default	active
1004 fddinet-default	active
1005 trnet-default	active

可以看到，由于域名和密码的不一致，所以导致 VTP 信息无法同步。此实验完成。

5.7 STP 基本配置

实验目的：
1. 掌握 STP 的基本配置。
2. 理解 STP 的选举机制。

实验拓扑：

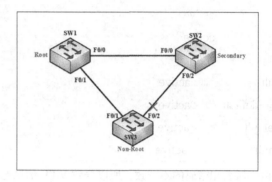

实验随手记：

_____
_____
_____
_____
_____
_____

实验原理：

1. STP 概述

STP（Spanning-tree Protocol，生成树协议）技术是用于局域网环境下交换环路的解决机制。当交换网络中有冗余链路并且出现广播包时，则网络会出现数据环路，如图 5-9 所示。数据环路会造成全网广播风暴、CAM 表抖动以及带宽资源占用等问题。STP 技术通过计算，在冗余链路中挑选出阻塞端口，执行数据阻塞，从而防止环路的形成。

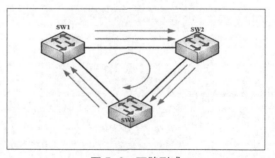

图 5-9 环路形成

【STP 协议】
从狭义上来讲，STP 协议指的是本节的生成树协议，从广义上来讲，STP 协议包括 STP、PVST、RSTP、MSTP 等生成树协议。STP 是所有生成树协议的基础和开始。

2. STP 术语

术语	描述
根桥	生成树的树根，整个树的参考点，用于维护整个生成树
备根桥	根桥的备份，接收并转发 BPDU 帧
非根桥	生成树中夹带阻塞端口的交换机
根端口	除了根桥，其他交换机必须具备根端口，用于接收 BPDU
指定端口	每根链路必须有指定端口，用于发送 BPDU
阻塞端口	每个冗余拓扑必须有阻塞端口，用于阻塞数据
BPDU	网桥协议数据单元，用于实现根的选举和根的维护
BID	网桥标识符，由网桥优先级和 MAC 组成，越小越优先
PID	接口标识符，由接口优先级和接口序号组成，越小越优先
COST	链路开销，衡量本地到根的距离，越小越优先

3. STP 选举

STP 生成树需要经过一次选举计算，最后才能得出阻塞端口，如图 5-10 所示。

①选举根桥：对比交换机的 BID。

②选举根端口：对比 COST、对比 BID、对比 PID。

③选举指定端口：对比 COST、对比 BID、对比 PID。

④选举阻塞端口：除了指定端口和根端口，剩下的便是阻塞端口。

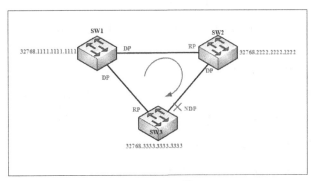

图 5-10　STP 防环

实验步骤：

1. 依据图中拓扑部署 STP 技术，其中 SW1 为 Root，SW2 为 Secondary，配置如下所示。

```
SW1(config)#spanning-tree vlan 1 priority 0
// 根据越小越优先原则，0 表示最优先，网桥优先级范围从 0 到 65535
SW2(config)#spanning-tree vlan 1 priority 4096
// 现在网桥优先级的修改必须是 4096 的倍数，有些老的机型则不受此限制
```

默认情况下，网桥优先级为 32768，则此时 SW3 上优先级为 32768。STP 根据越小越优先原则进行根桥选举。

2. 查看网桥选举情况。

SW1 上网桥选举情况，如下所示。

【BPDU】
生成树协议交互的信息，类似路由协议的分组更新信息，夹带选举根桥所需要的开销、BID、PID 等信息。

【BID】
Bridge Identifier，网桥标识符，类似路由标识符 RID，用于唯一标识网络中的交换机，由优先级和 MAC 地址组成。默认网桥优先级为 32768，范围为 0 ~ 65535，越小越优先。

【PID】
由接口优先级和接口序号组成，接口优先级范围为 0 ~ 255，默认为 128，越小越优先。

【COST】
开销值，根据带宽不同而不同，例如 10Mbit 为 100，100Mbit 为 19，1Gbit 为 4，10Gbit 为 2。

```
SW1#show spanning-tree vlan 1 brief
// 查看关于此 VLAN 的 STP 简要信息
VLAN1
  Spanning tree enabled protocol ieee
  Root ID    Priority    0
             Address     cc05.23d8.0000
             This bridge is the root
// 表示本地便是根桥，一般只有在根桥上出现
             Hello Time  2 sec  Max Age 20 sec  Forward Delay 15 sec

  Bridge ID  Priority    0
             Address     cc05.23d8.0000
             Hello Time  2 sec  Max Age 20 sec  Forward Delay 15 sec
             Aging Time 300

Interface                        Designated
Name         Port ID Prio Cost Sts Cost Bridge ID       Port ID
---------------------- ------- ---- ----- --- ----- -----------
FastEthernet0/0 128.1   128  19  FWD  0   0  cc05.23d8.0000 128.1
FastEthernet0/1 128.2   128  19  FWD  0   0  cc05.23d8.0000 128.2
```

从上面可以看出，SW1 已经被选举为根桥，并且所有的接口处于转发状态。SW2 上网桥选举情况，如下所示。

```
SW2#show spanning-tree vlan 1 brief
// 查看关于此 VLAN 的 STP 简要信息
VLAN1
  Spanning tree enabled protocol ieee
   Root ID   Priority    0
             Address     cc05.23d8.0000
             Cost        19
             Port        1 (FastEthernet0/0)
             Hello Time  2 sec  Max Age 20 sec  Forward Delay 15 sec

  Bridge ID  Priority    4096
             Address     cc06.23d8.0000
             Hello Time  2 sec  Max Age 20 sec  Forward Delay 15 sec
```

```
                     Aging Time  300
Interface                         Designated
Name        Port ID  Prio Cost  Sts  Cost  Bridge ID       Port ID
-------------------- -------- ---- ----- --- ----- -----------
FastEthernet0/  128.1  128  19  FWD  0    0      cc05.23d8.0000 128.1
FastEthernet0/2 128.3  128  19  FWD  19  4096   cc06.23d8.0000 128.3
```

SW3 上网桥选举情况，如下所示。

```
SW3#show spanning-tree vlan 1 brief

VLAN1
  Spanning tree enabled protocol ieee
  Root ID    Priority    0
             Address     cc05.23d8.0000
             Cost        19
             Port        2 (FastEthernet0/1)
             Hello Time  2 sec  Max Age 20 sec  Forward Delay 15 sec

  Bridge ID  Priority    32768
             Address     cc07.23d8.0000
             Hello Time  2 sec  Max Age 20 sec  Forward

  Delay 15 sec
             Aging Time  300

Interface                        Designated
Name       Port ID  Prio Cost  Sts  Cost  Bridge ID       Port ID
-------------------- -------- ---- ----- --- ----- -----------
FastEthernet0/1 128.2  128  19  FWD  0    0 cc05.23d8.0000 128.2
FastEthernet0/2 128.3  128  19  BLK  19  4096 cc06.23d8.0000 128.3
// 为了防环，把 SW3 的 Fa0/2 端口阻塞了
```

SW2 的所有接口处于转发状态，SW3 的优先级为 32768，并且 F0/2 接口处于阻塞状态，用于防止环路。

3. 工程用法。一般项目环境习惯直接指定网桥角色，不需要指定具体优先级，配置如下所示。

```
SW1(config)#spanning-tree vlan 1 root primary
// 此配置原理是交换机检测全局的优先级，然后将本地的优先级调整到适当的位置，
以达到目的。一般此命令会转换为具体数值，并可以通过 show run 查看到
SW2(config)#spanning-tree vlan 1 root secondary
```

4. 扩展。若需要修改 RP 和 DP，可以在接口下修改链路开销值和端口优先级，配置如下所示。

```
SW1(config)#int f0/0
SW1(config-if)#spanning-tree cost ?
<1-65535>  Change an interface's spanning tree path cost
SW1(config-if)#spanning-tree port-priority ?
<0-255>  Change an interface's spanning tree priority
```

此实验完成。

5.8 STP 进阶配置

实验目的：

1. 掌握 Portfast、Uplinkfast、Backbonefast 的配置和原理。
2. 理解 STP 的链路收敛。

实验拓扑：

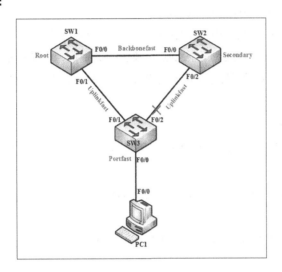

实验随手记：

___

___

___

___

___

实验原理：

1. STP 端口状态

STP 的端口状态需要经过一个状态机的过程，如图 5-11 所示。

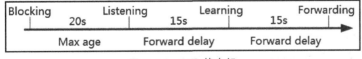

图 5-11　STP 状态机

Blocking（阻塞状态）：执行阻塞端口。

Listening（侦听状态）：执行端口选举。

Learning（学习状态）：执行地址学习。

Forwarding（转发状态）：执行数据转发。

2. STP 链路收敛

链路收敛是指网络发生拓扑变更到链路重新回到稳定状态的整个过程，由于 STP 存在端口状态机，所以当链路发生变动时，链路收敛需要经过一段时间。直连链路收敛需要经过 30s，而间接链路收敛需要经过 50s。

3. STP 增强特性

由于 STP 默认的收敛速度非常缓慢，Cisco 提出 3 个私有的 STP 增强特性，即 Portfast、uplinkfast、backbonefast。

Portfast：端口加速，用于加速用户接入时间，能够加速 30s。

Uplinkfast：上联加速，用于加速直接链路收敛时间，能够加速 30s。

Backbonefast：骨干加速，用于加速间接链路收敛时间，能够加速 20s。

实验步骤：

1. 依据图中拓扑部署 STP 技术，其中 SW1 为 Root，SW2 为 Secondary，配置如下所示。

```
SW1(config)#spanning-tree vlan 1 root primary
// 设置 SW1 为 vlan 1 的 Spanning-tree 的主根树
SW2(config)#spanning-tree vlan 1 root secondary
// 设置 SW2 为 vlan 1 的 Spanning-tree 的次根
```

2. 理解 STP 链路加速特性。默认情况下，STP 的链路收敛至少需要 30s 或者 50s，其中直接链路收敛 30s，间接链路收敛 50s。由于收敛速度非常缓慢，Cisco 引入三种私有特性，portfast/uplinkfast/backbonefast，各自功能如下所示。

① Portfast：端口加速，主要用于接入层交换机的接入接口，用于减少用户接入网络的时间，可加速 30s。只需要在接入层交换机上部署。

② Uplinkfast：上联加速，主要用于接入层交换机的上联链路，当上联链路出现故障后，备用链路能够快速切换，可加速 30s，此特性用于加速 STP 的直接收敛。只需要在接入层交换机上部署。

③ Backbonefast：骨干加速，主要用于汇聚层交换机的骨干链路，当骨干链路出现故障后，阻塞端口能够快速切换，可加速 20s，此特性用于加速 STP 的间接收敛。需要在所有交换机上部署。

3. 部署 STP 链路加速特性，配置如下所示。

```
Portfast=>
SW3(config)#int f0/0
SW3(config-if)#spanning-tree portfast
Uplinkfast=>
SW3(config)#spanning-tree uplinkfast
Backbonefast=>
SW1(config)#spanning-tree backbonefast
SW2(config)#spanning-tree backbonefast
```

```
SW3(config)#spanning-tree backbonefast
```

4. 验证 STP 加速特性。

①验证 Portfast 特性，如下所示。

```
SW3#debug spanning-tree events
SW3(config)#int f0/0
SW3(config-if)#sh
// 把 SW3 的 f0/0 端口关闭
*Mar 1 01:51:10.243: STP: VLAN1 Fa0/0 -> blocking
SW3(config-if)#no sh
// 再打开
*Mar 1 01:51:22.603: STP: VLAN1 Fa0/0 ->jump to forwarding from blocking
// 立刻由阻塞状态转变成了转发状态。
```

②验证 Uplinkfast 特性，如下所示。

```
SW3#debug spanning-tree events
SW3(config)#int f0/1
SW3(config-if)#sh
*Mar  1 01:57:49.011: STP: VLAN1 Fa0/1 -> blocking
*Mar  1 01:57:49.011: STP: VLAN1 new root port Fa0/2, cost 3038
*Mar  1 01:57:49.011: STP FAST: UPLINKFAST: make_forwarding on VLAN1
FastEthernet0/2 root port id new: 128.3 prev: 128.2

*Mar  1 01:57:49.011: %SPANTREE_FAST-7-PORT_FWD_UPLINK: VLAN1
FastEthernet0/2 moved to Forwarding (UplinkFast).
// 加速备用链路的收敛过程
```

③验证 Backbonefast 特性，如下所示。

```
SW3#debug spanning-tree events
SW3#debug spanning-tree backbonefast
SW2(config)#int f0/0
SW2(config-if)#sh
*Mar 1 02:10:34.987: STP: VLAN1 Fa0/2 ->listening
*Mar 1 02:10:35.963: STP: VLAN1 heard root 16384-cc06.23d8.0000 on Fa0/2
*Mar1 02:10:35.963: current Root has 0-cc05.23d8.0000
*Mar1 02:10:35.963: STP FAST: received inferior BPDU on VLAN1
FastEthernet0/2.
*Mar 1 02:10:49.991: STP: VLAN1 Fa0/2 ->learning
*Mar 1 02:11:05.011: STP: VLAN1 Fa0/2 ->forwarding
```

通过 debug 命令可以看到详细的状态变化。

5.9 PVST 基本配置

实验目的：
1. 掌握 PVST 的基本配置。
2. 理解 PVST 的负载均衡原理。

实验拓扑：

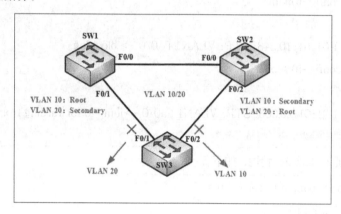

实验随手记：

_____
_____
_____
_____
_____

实验原理：

PVST<Per-Vlan STP，基于 VLAN 的 STP> 是 Cisco 私有的生成树协议，PVST 技术是在 STP 的基础上进行开发的。原有 STP 具有 CST（Common Spanning-tree，公共生成树）特性，即所有的 VLAN 默认共享一棵树，所有 VLAN 有统一的根桥和统一的阻塞端口，所有 VLAN 的流量只会往某条链路走，而这样的话，流量无法实现负载均衡，从而导致网络资源浪费。PVST 技术可以实现不同 VLAN 不同树的计算，不同 VLAN 的生成树相互独立，不同 VLAN 有不同的阻塞端口，因而通过参数调整，可以有效实现流量负载均衡。目前，PVST 是一些 Cisco 交换机如 C3550 和 C3560 的默认生成树协议。

实验步骤：

1. 依据图中拓扑部署 PVST 技术，其中 SW1 为 VLAN10 的 Root，VLAN20 的 Secondary；SW2 为 VLAN20 的 Root，VLAN10 的 Secondary，配置如下所示。

```
SW1(config)#spanning-tree vlan 10 root primary
// 设置 SW1 为 vlan 10 的主根
SW1(config)#spanning-tree vlan 20 root secondary
// 设置 SW1 为 vlan 20 的次根
SW2(config)#spanning-tree vlan 20 root primary
// 设置 SW2 为 vlan 20 的主根
SW2(config)#spanning-tree vlan 10 root secondary
// 设置 SW2 为 vlan 10 的次根
```

2. 查看 STP 状态。

SW1 上 STP 状态如下所示。

```
SW1#show spanning-tree vlan 10 brief
        // 查看 vlan 10 的 Spanning-tree 的主要情况
VLAN10
  Spanning tree enabled protocol ieee
  Root ID    Priority    8192
             Address    cc05.0810.0001
    This bridge is the root
//SW1 是 VLAN 10 的 root 根
          Hello Time   2 sec  Max Age 20 sec  Forward
Delay 15 sec

  Bridge ID  Priority    8192
             Address    cc05.0810.0001
             Hello Time   2 sec  Max Age 20 sec  Forward Delay 15 sec
             Aging Time 300


Interface                       Designated
Name              Port ID Prio Cost Sts Cost Bridge ID      Port ID
-------------------- ------- ---- ----- --- ----- ----------
FastEthernet0/0    128.1   128   19 FWD    0 8192 cc05.0810.0001 128.1
FastEthernet0/1    128.2   128   19 FWD    0 8192 cc05.0810.0001 128.2
```

可以看到，SW1 已经成为 VLAN 10 的主根。

SW2 上 STP 状态如下所示。

```
SW2#show spanning-tree vlan 20 brief

VLAN20
```

```
Spanning tree enabled protocol ieee
Root ID    Priority    8192
           Address     cc06.0810.0002
           This bridge is the root
           Hello Time   2 sec  Max Age 20 sec  Forward Delay 15 sec

Bridge ID  Priority    8192
           Address     cc06.0810.0002
           Hello Time   2 sec  Max Age 20 sec  Forward Delay 15 sec
           Aging Time 300

Interface                         Designated
Name              Port ID Prio Cost Sts Cost  Bridge ID         Port ID
--------------------------------- ------- ---- ----- --- ----- -----------
FastEthernet0/0   128.1   128   19 FWD    0 8192 cc06.0810.0002 128.1
FastEthernet0/2   128.3   128   19 FWD    0 8192 cc06.0810.0002 128.3
```

可以看到，SW2 已经成为 VLAN 20 的主根。

SW3 上 STP 状态如下所示。

```
SW3#show spanning-tree vlan 10 brief

VLAN10
 Spanning tree enabled protocol ieee
 Root ID    Priority    8192
            Address     cc05.0810.0001
            Cost        19
            Port        2 (FastEthernet0/1)
            Hello Time   2 sec  Max Age 20 sec  Forward Delay 15 sec

 Bridge ID  Priority    32768
            Address     cc07.0810.0001
            Hello Time   2 sec  Max Age 20 sec  Forward Delay 15 sec
            Aging Time 300

Interface                         Designated
Name           Port ID Prio Cost Sts Cost  Bridge ID   Port ID
```

```
--------------------- ------- ---- ----- --- ----- ----------
FastEthernet0/1 128.2   128    19  FWD    0  8192 cc05.0810.0001 128.2
FastEthernet0/2 128.3   128    19  BLK   19 16384 cc06.0810.0001 128.3
```

```
SW3#show spanning-tree vlan 20 brief

VLAN20
  Spanning tree enabled protocol ieee
  Root ID   Priority    8192
            Address     cc06.0810.0002
            Cost        19
            Port        3 (FastEthernet0/2)
            Hello Time  2 sec  Max Age 20 sec  Forward Delay 15 sec

  Bridge ID Priority    32768
            Address     cc07.0810.0002
            Hello Time  2 sec  Max Age 20 sec  Forward Delay 15 sec
            Aging Time 300

Interface                     Designated
Name           Port ID Prio  Cost Sts  Cost Bridge ID  Port ID
--------------------- ------- ---- ----- --- ----- ----------
FastEthernet0/1 128.2   128    19  BLK   19 16384 cc05.0810.0002 128.2
FastEthernet0/2 128.3   128    19  FWD    0  8192 cc06.0810.0002 128.3
```

从上面可以看到，SW3 在 VLAN10 所在生成树中阻塞 F0/2，在 VLAN20 所在生成树中阻塞 F0/1，通过此方法可以实现负载均衡。此实验完成。

5.10 单臂路由

实验目的:
1. 掌握单臂路由实现 VLAN 间通信的基本配置。
2. 理解单臂路由的实现原理。

实验拓扑:

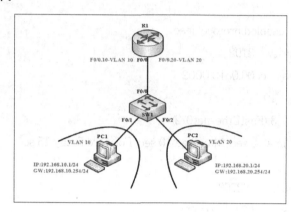

实验随手记:

_____
_____
_____
_____
_____

实验原理:

1. 单臂路由概述

单臂路由技术可以实现局域网不同 VLAN 之间的数据互访,通过在路由器物理接口划分不同的逻辑子接口实现。逻辑子接口映射到不同的 VLAN,承载不同网段的流量,逻辑子接口具备独立的 IP 地址,并且 IP 地址一般为 VLAN 的网关地址。

2. 单臂路由特征

单臂路由虽然能实现 VLAN 间通信,但是随着 VLAN 数目的增加,路由器和交换机之间承载的压力非常大,容易造成单链路故障。所以在目前的网络拓扑设计中,单臂路由已经慢慢淡出。

【单臂路由】
一种用于实现 VLAN 间通信的技术,通过划分子接口实现。但是由于其拓展性不好,取而代之的是通过三层交换机来实现。

实验步骤:

1. 依据图中拓扑,在 SW1 上创建 VLAN,配置如下所示。

SW1#vlan database
// 一般 3640 或者 3725 等系列路由器的交换模块需要进入 VLAN 数据库模式进行操作

```
SW1(vlan)#vlan 10 name VLAN_10
// 创建 VLAN 并定义 VLAN 名字，一般不同的应用或者部门采用不同的命名，易于
管理
SW1(vlan)#vlan 20 name VLAN_20
SW1(vlan)#exit
```

将接口划入对应 VLAN，配置如下所示。

```
SW1(config)#int f0/1
SW1(config-if)#switchport mode access
// 将接口模式修改为接入模式，此模式一般用于接入终端主机
SW1(config-if)#switchport access vlan 10
// 将接口放入 VLAN 10
SW1(config-if)#int f0/2
SW1(config-if)#switchport mode access
SW1(config-if)#switchport access vlan 20
SW1(config-if)#exit
```

2. 根据拓扑为不同 PC 配置 IP 地址信息，如下所示。

```
PC1(config)#no ip routing
// 关闭三层路由功能，模拟主机
PC1(config)#int f0/0
PC2(config-if)#no sh
PC1(config-if)#ip address 192.168.10.1 255.255.255.0
PC1(config-if)#exit
PC1(config)#ip default-gateway 192.168.10.254
// 定义默认网关
PC2(config)#no ip routing
PC2(config)#int f0/0
PC2(config-if)#no sh
PC2(config-if)#ip address 192.168.20.1 255.255.255.0
PC2(config-if)#exit
PC2(config)#ip default-gateway 192.168.20.254
```

3. 将 SW1 与 R1 相连链路设置为 Trunk 链路，用于承载不同 VLAN 的流量，
配置如下所示。

```
SW1(config)#int f0/0
SW1(config-if)#switchport trunk encapsulation dot1q
// Trunk 有两种封装标准，一种是 Cisco 私有的 ISL，一种为行业标准 802.1Q，一般
采用 802.1Q 实现封装
SW1(config-if)#switchport mode trunk
```

// 将接口模式定义为 trunk 模式,交换机相连的接口一般采用 trunk 模式,用于承载不同 VLAN 的流量。

SW1(config-if)#exit

4. 在 R1 上部署单臂路由技术,实现 VLAN 间通信,如下所示。

R1(config)#int f0/0

R1(config-if)#no shutdown

R1(config-if)#exit

R1(config)#int f0/0.10

// 开启子接口

R1(config-subif)#encapsulation dot1Q 10

// 将此子接口封装 dot1q,并划入 VLAN10。后面那个 10 是划入 VLAN 10 的意思

R1(config-subif)#ip address 192.168.10.254 255.255.255.0

R1(config-subif)#exit

R1(config)#int f0/0.20

R1(config-subif)#encapsulation dot1Q 20

// 将此子接口封装 dot1q,并划入 VLAN20

R1(config-subif)#ip address 192.168.20.254 255.255.255.0

R1(config-subif)#exit

5. 测试单臂路由。

①测试主机到网关的连通性,如下所示。

PC1#ping 192.168.10.254

Type escape sequence to abort.

Sending 5, 100-byte ICMP Echos to 192.168.10.254, timeout is 2 seconds:

!!!!!

Success rate is 100 percent (5/5), round-trip min/avg/max = 20/227/1036 ms

②测试主机与主机的连通性,如下所示。

PC1#ping 192.168.20.1

Type escape sequence to abort.

Sending 5, 100-byte ICMP Echos to 192.168.20.1, timeout is 2 seconds:

.!!!!

Success rate is 80 percent (4/5), round-trip min/avg/max = 24/48/84 ms

从上面可以看到,通过部署单臂路由技术,VLAN 间的主机能够相互通信。此实验完成。

5.11 三层交换机

实验目的：
1. 掌握三层交换机实现 VLAN 间通信的基本配置。
2. 理解三层交换机的实现原理。

实验拓扑：

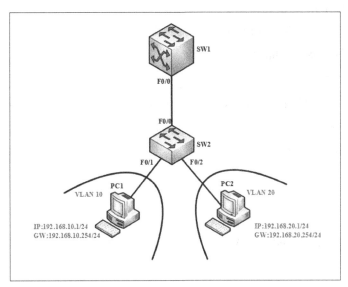

实验随手记：

_____
_____
_____
_____
_____

实验原理：

采用三层交换机实现 VLAN 通信是目前最为主流的局域网拓扑架构。三层交换机融合了二层交换和三层路由的功能，而且采用专用的 ASIC 硬件芯片卡进行数据转发，在性能上更胜一筹。三层交换机的快速以太口数量比一般的路由器多，在端口成本上也有优势。三层交换机通过 SVI<Switching Virtual Interface，交换虚拟接口>实现不同 VLAN 的通信。

【三层交换机】
三层交换机或者多层交换机是目前局域网的主流产品，一般设计在企业网或者校园网的汇聚层，用于实现 VLAN 间通信。

实验步骤：
1. 依据图中拓扑，在 SW1 和 SW2 之间部署 Trunk 链路，配置如下所示。

```
SW1(config)#int f0/0
SW1(config-if)#switchport trunk encapsulation dot1q
```

// Trunk 有两种封装标准，一种是 Cisco 私有的 ISL，一种是行业标准 802.1Q，一般采用 802.1Q 实现封装

SW1(config-if)#switchport mode trunk

// 将接口模式定义为 trunk 模式，交换机相连的接口一般采用 trunk 模式，用于承载不同 VLAN 的流量

SW1(config-if)#exit

SW2(config)#int f0/0

SW2(config-if)#switchport trunk encapsulation dot1q

SW2(config-if)#switchport mode trunk

SW2(config-if)#exit

2. 部署 VTP 技术，SW1 为 Server，SW2 为 Client，实现 VLAN 同步，如下所示。

SW1#vlan database

SW1(vlan)#vtp server

// 定义 VTP 模式，全局模式方式：

(config)vtp mode server/client

SW1(vlan)#vtp domain CCNA

// 定义 VTP 管理域，与全局模式配置方法一致

SW1(vlan)#vtp password Cisco

// 定义 VTP 密码，实现 VTP 安全，与全局模式配置方法一致

// 域名和密码需要一致

SW1(vlan)#vlan 10

SW1(vlan)#vlan 20

SW1(vlan)#exit

SW2#vlan database

SW2(vlan)#vtp client

SW2(vlan)#vtp domain CCNA

SW2(vlan)#vtp password cisco

SW2(vlan)#exit

3. 确认 VTP 信息同步并将接口放入相应 VLAN，如下所示。

SW2#show vlan-switch brief

// 查看该交换机 vlan 的主要情况

VLAN Name   Status   Ports

1 default	active	Fa0/1,Fa0/2,Fa0/3,Fa0/4,Fa0/5,Fa0/6
		Fa0/7, Fa0/8, Fa0/9, Fa0/10
		Fa0/11, Fa0/12, Fa0/13, Fa0/14
		Fa0/15
10 VLAN0010	active	
20 VLAN0020	active	

//VLAN 10 和 VLAN 20 均被成功添加

1002 fddi-default	active
1003 token-ring-default	active
1004 fddinet-default	active
1005 trnet-default	active

可以看到，SW2 已经学习到 VLAN10 和 VLAN20，将接口划入，配置如下所示。

```
SW2(config)#int f0/1
SW2(config-if)#switchport mode access
// 将接口模式修改为接入模式，此模式一般用于接入终端主机
SW2(config-if)#switchport access vlan 10
// 将接口放入 VLAN 10
SW2(config-if)#int f0/2
SW2(config-if)#switchport mode access
SW2(config-if)#switchport access vlan 20
SW2(config-if)#exit
```

4. 根据拓扑为不同 PC 配置 IP 地址信息，如下所示。

```
PC1(config)#no ip routing
// 关闭三层路由功能，模拟主机
PC1(config)#int f0/0
PC2(config-if)#no sh
PC1(config-if)#ip address 192.168.10.1 255.255.255.0
PC1(config-if)#exit
PC1(config)#ip default-gateway 192.168.10.254
// 定义默认网关
PC2(config)#no ip routing
PC2(config)#int f0/0
PC2(config-if)#no sh
PC2(config-if)#ip address 192.168.20.1 255.255.255.0
```

PC2(config-if)#exit

PC2(config)#ip default-gateway 192.168.20.254

5. 在三层交换机上开启三层路由功能，并为不同 VLAN 定义不同 IP，如下所示。

SW1(config)#ip routing

// 开启三层路由功能，默认情况下，三层交换机关闭路由功能，若要实现 VLAN 间通信，需要开启。

SW1(config)#int vlan 10

// 进入到 VLAN 10 的接口模式

SW1(config-if)#ip address 192.168.10.254 255.255.255.0

// 为 VLAN 10 配置 IP 地址，我们将配置了 IP 地址的 VLAN 叫做 SVI 接口，即 Swithcing Virtual Interface，虚拟交换接口，它是一种三层逻辑口。一般在 VLAN 上面配置 IP 网关，然后主机 IP 地址指向三层交换机，三层交换机通过多个 SVI 接口实现 VLAN 间通信

SW1(config-if)#exit

SW1(config)#int vlan 20

SW1(config-if)#ip address 192.168.20.254 255.255.255.0

SW1(config-if)#exit

6. 测试三层交换。

①测试主机到网关的连通性，如下所示。

PC1#ping 192.168.10.254

Type escape sequence to abort.

Sending 5, 100-byte ICMP Echos to 192.168.10.254, timeout is 2 seconds:

!!!!!

Success rate is 100 percent (5/5), round-trip min/avg/max = 20/227/1036 ms

②测试主机与主机的连通性，如下所示。

PC1#ping 192.168.20.1

Type escape sequence to abort.

Sending 5, 100-byte ICMP Echos to 192.168.20.1, timeout is 2 seconds:

.!!!!

Success rate is 80 percent (4/5), round-trip min/avg/max = 24/48/84 ms

从上面可以看到，通过部署三层交换技术，VLAN 间的主机能够相互通信。此实验完成。

5.12 DHCP 基本配置

实验目的：
1. 掌握 DHCP 基本配置实现地址自动分配。
2. 理解 DHCP 的实现原理。

实验拓扑：

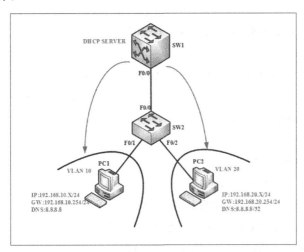

实验随手记：

_____
_____
_____
_____
_____
_____

实验原理：

1. DHCP 概述

DHCP（Dynamic Host Configuration Protocol，动态主机配置协议）技术可以实现对终端设备 IP 信息的动态分配和地址的集中管理。DHCP 技术有 DHCP Server 和 DHCP Client 两种角色，由 Client 端向 Server 端发起地址请求，Server 从地址池中向 Client 分配地址。

2. DHCP 分组

DHCP Client 与 Server 获取地址的过程涉及 4 个分组，具体如下。

Discover 分组：客户端向网络发起广播发现包。

Offer 分组：服务端收到客户的发现包后，返回提供包。

Request 分组：客户端向服务端正式发起地址请求。

Ack 分组：服务端正式向客户端提供地址。

实验步骤：

1. 依据图中拓扑，在 SW1 上架设 DHCP 服务，为不同 VLAN 的主机动态提供 IP 信息，配置如下所示。

```
SW1(config)#ip dhcp pool VLAN10
// 定义 DHCP 地址池，地址池命名尽量与所分配的部门或者 VLAN 挂钩，方便管理
SW1(dhcp-config)#network 192.168.10.0 255.255.255.0
// 定义地址池的网段，需要定义子网掩码
SW1(dhcp-config)#default-router 192.168.10.254
// 定义默认网关
SW1(dhcp-config)#dns-server 8.8.8.8 114.114.114.114
// 定义 DNS 服务器，可以定义多个，用于实现主备，一般第一个 DNS 地址为主 DNS
SW1(dhcp-config)#lease 7
// 定义地址释放时间，这里是 7 天
SW1(dhcp-config)#exit
SW1(config)#ip dhcp pool VLAN20
SW1(dhcp-config)#network 192.168.20.0 255.255.255.0
SW1(dhcp-config)#default-router 192.168.20.254
SW1(dhcp-config)#dns-server 8.8.8.8 114.114.114.114
SW1(dhcp-config)#lease 7
SW1(dhcp-config)#exit
```

2. 在 DHCP Client 将 IP 设置为动态学习方式，如下所示。

```
PC1(config)#int f0/0
PC1(config-if)#ip address dhcp
// 设置接口通过 DHCP 方式动态获取地址
PC1(config-if)#exit
PC2(config)#int f0/0
PC2(config-if)#ip address dhcp
PC2(config-if)#exit
```

3. 检测 DHCP 分配情况。

① 在 DHCP Client 上查看地址是否学到，如下所示。

```
PC1#show ip int brief
Interface    IP-Address    OK? Method Status    Protocol
FastEthernet0/0 192.168.10.1 YES DHCP   up       up
// 地址获取方式从原来的 manual 变成 DHCP
PC2#show ip int brief
Interface    IP-Address    OK? Method Status    Protocol
FastEthernet0/0 192.168.20.1 YES DHCP   up       up
```

从上面可以看出，PC 的地址已经学到，并且地址学习方式从手工方式变成 DHCP 方式。

② 在 DHCP Server 上查看 DHCP 地址池的分配情况，如下所示。

```
SW1#show ip dhcp pool
// 查看 DHCP 地址池的信息
Pool VLAN10 :
 Utilization mark (high/low)    : 100 / 0
 Subnet size (first/next)       : 0 / 0
 Total addresses                : 254
// 此地址池总共有多少个地址
 Leased addresses               : 1
// 已经分配出的地址数目
 Pending event                  : none
 1 subnet is currently in the pool :
 Current index     IP address range              Leased addresses
 192.168.10.2      192.168.10.1 – 192.168.10.254         1


Pool VLAN20 :
 Utilization mark (high/low)    : 100 / 0

 Subnet size (first/next)       : 0 / 0
 Total addresses                : 254
 Leased addresses               : 1
 Pending event                  : none
 1 subnet is currently in the pool :
 Current index     IP address range              Leased addresses
 192.168.20.2      192.168.20.1 – 192.168.20.254         1
```

③ 查看 DHCP 地址分配和主机情况，如下所示。

```
SW1#show ip dhcp binding
// 查看主机地址获取信息，包括 IP、MAC、租约等
Bindings from all pools not associated with VRF:
IP address        Client-ID/              Lease expiration       Type
                  Hardware address/
                  User name
192.168.10.1      0063.6973.636f.2d63.    Mar 08 2002 12:37 AM   Automatic
                  6330.352e.3330.3963.
                  2e30.3030.302d.4661.
```

	302f.30		
192.168.20.1	0063.6973.636f.2d63.	Mar 08 2002 12:38 AM	Automatic
	6330.362e.3330.3963.		
	2e30.3030.302d.4661.		
	302f.30		

  从上面的结果可以看到，通过部署 DHCP 技术，主机可以更方便地接入网络。尤其是移动互联网时代，DHCP 技术越发彰显其重要性，各种终端如 PC、手机、平板需要更方便地连入网络。因此，除了常规的企业网络，在现在的公共 WIFI 场所如公交、地铁、咖啡馆等，DHCP 技术的应用非常广泛。此实验完成。

5.13　L2 Etherchannel 基本配置

实验目的：
1. 掌握 L2 Etherchannel 的基本配置。
2. 理解 Etherchannel 的协商协议。

实验拓扑：

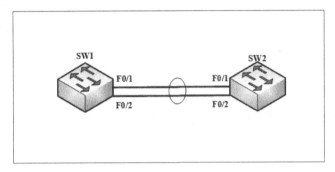

实验随手记：

_____
_____
_____
_____
_____
_____

实验原理：

1. 以太通道概述

Etherchannel（以太通道）技术是运行在局域网内部用于将多根物理链路捆绑在一起的技术，可以有效提高链路带宽和增强网络冗余性。以太通道技术分为 L2 和 L3 两种，L2 一般在交换技术下执行，L3 在路由环境下执行。

2. 以太通道协商

以太通道的建立可以直接通过强制方式，在交换机之间直接启用，也可以通过动态协商协议如 PAGP 或 LACP 进行协商建立。PAGP 协议是思科私有协议，采用 Desirable 和 Auto 模式进行组合协商；而 LACP 是业界标准协议 IEEE 802.3ad，采用 Active 和 Passive 模式进行组合协商。协商结果与 DTP 类似。

实验步骤：

1. 依据图中拓扑，在 SW1 和 SW2 上部署 L2 Etherchannel，实现链路捆绑并提供链路带宽，可以通过 3 种方式部署，具体如下。

①强制方式：直接绑定，无需任何协商。

【以太通道】以太通道技术分为2层和3层以太通道。2层以太通道无法配置 IP 地址，3层以太通道可以。

② Pagp 方式：通过 Cisco 私有协议 Pagp 进行协商。

③ Lacp 方式：通过行业标准协议 Lacp 进行协商。

2. 依据强制方式部署 L2 Etherchannel。

SW1 上配置如下所示。

SW1(config)#int range f0/1－2

SW1(config-if-range)#switchport trunk encapsulation dot1q

// Trunk 有两种封装标准，一种是 Cisco 私有的 ISL，另一种是行业标准 802.1Q，一般采用 802.1Q 实现封装

SW1(config-if)#switchport mode trunk

// 将接口模式定义为 trunk 模式，交换机相连的接口一般采用 trunk 模式，用于承载不同 VLAN 的流量。

SW1(config-if-range)#channel-group 1 mode on

// 将 F0/1 和 F0/2 放入 group1，其中 group 为本地标识，用于区分本地多个捆绑组

SW1(config-if-range)#exit

SW2 上配置如下所示。

SW2(config)#int range f0/1－2

SW2(config-if-range)#switchport trunk encapsulation dot1q

SW2(config-if-range)#switchport mode trunk

SW2(config-if-range)#channel-group 1 mode on

SW2(config-if-range)#exit

此时查看以太通道信息，如下所示。

SW1#show etherchannel summary

// 显示端口捆绑的简要信息

Flags: D － down   P － in port-channel

　　　　I － stand-alone s － suspended

　　　　H － Hot-standby (LACP only)

　　　　R － Layer3   S － Layer2

　　　　U － in use   f － failed to allocate aggregator

　　　　u － unsuitable for bundling

　　　　w － waiting to be aggregated

　　　　d － default port

Number of channel-groups in use: 1

Number of aggregators: 1

Group Port-channel Protocol   Ports

```
-------+-------------+-----------+-------------------------
1       Po1(SU)       -          Fa0/1(P) Fa0/2(P)
```

此时以太通道已经成功建立，查看以太通道物理信息，如下所示。

```
SW1#show interfaces port-channel 1
```
// 显示聚会后的 port-channel 1 信息

Port-channel1 is up, line protocol is up
  Hardware is EtherChannel, address is cc03.0a28.f000 (bia cc03.0a28.f000)
  MTU 1500 bytes, BW 200000 Kbit, DLY 1000 usec,
// 带宽加倍了
    reliability 255/255, txload 1/255, rxload 1/255
  Encapsulation ARPA, loopback not set
  Keepalive set (10 sec)
  Full-duplex, 100Mb/s
  Members in this channel: Fa0/0 Fa0/1
  ARP type: ARPA, ARP Timeout 04:00:00
  Last input 00:00:01, output never, output hang never
  Last clearing of "show interface" counters never
  Input queue: 0/75/0/0 (size/max/drops/flushes); Total output drops: 0
  Queueing strategy: fifo
  Output queue: 0/40 (size/max)
  5 minute input rate 0 bits/sec, 0 packets/sec
  5 minute output rate 0 bits/sec, 0 packets/sec
     0 packets input, 0 bytes, 0 no buffer
     Received 0 broadcasts, 0 runts, 0 giants, 0 throttles
     0 input errors, 0 CRC, 0 frame, 0 overrun, 0 ignored
     0 input packets with dribble condition detected
     0 packets output, 0 bytes, 0 underruns
     0 output errors, 0 collisions, 1 interface resets
     0 babbles, 0 late collision, 0 deferred
     0 lost carrier, 0 no carrier
     0 output buffer failures, 0 output buffers swapped out

可以看到以太通道的带宽为 200Mbit。

3. 采用 Pagp 方式进行协商绑定。

SW1 上配置如下所示。

```
SW1(config)#int range f0/1 - 2
```

```
SW1(config-if-range)#switchport trunk encapsulation dot1q
SW1(config-if-range)#switchport mode trunk
SW1(config-if-range)#channel-protocol pagp
// 可以采用 PAGP 或者 LACP 协议
SW1(config-if-range)#channel-group 1 mode desirable
// PAGP 有 desirable 和 auto 模式，LACP 有 active 和 passive 模式
```

SW2 上配置如下所示。

```
SW2(config)#int range f0/1 - 2
SW2(config-if-range)#switchport trunk encapsulation dot1q
SW2(config-if-range)#switchport mode trunk
SW2(config-if-range)#channel-protocol pagp
SW2(config-if-range)#channel-group 1 mode auto
```

查看以太通道状态，如下所示。

```
SW1#show etherchannel summary
Flags:  D - down        P - in port-channel
        I - stand-alone s - suspended
        H - Hot-standby (LACP only)
        R - Layer3      S - Layer2
        U - in use      f - failed to allocate aggregator

        u - unsuitable for bundling
        w - waiting to be aggregated
        d - default port


Number of channel-groups in use: 1
Number of aggregators:           1

Group  Port-channel  Protocol    Ports
------+-------------+-----------+-----------------------

1      Po1(SU)        PAgP    Fa0/1(P) Fa0/2(P)
```

可以看到，通过 Pagp 协议实现了 L2 Etherchannel 的捆绑，同样也可以采用 Lacp 实现，此处略过。此实验完成。

第 6 章　广域网技术

本章主要学习广域网技术，研究不同局域网分支站点的互联。我们将学习到广域网技术中各种通信协议包括 HDLC、PPP 和 Frame-Relay。通过本章学习，我们将掌握广域网远程连接的原理和实现。以下是本章导航图：

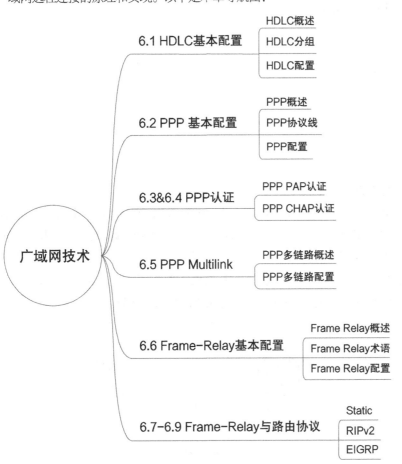

【广域网】
WAN<Wide Area Network> 广域网是在局域网技术的基础上发展开来的，用于实现对不同局域网之间的连接。广域网通过路由器进行远程相连，通过广域网协议传输数据包。

6.1 HDLC 基本配置

实验目的：

1. 掌握 HDLC 基本配置。
2. 理解 HDLC 封装格式。

实验拓扑：

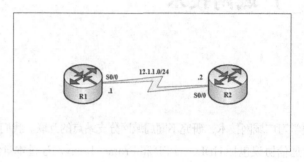

实验随手记：

_____
_____
_____
_____
_____

实验原理：

HDLC（High-Level Data Link Control，高级链路数据控制）协议是广域网点对点串行链路上的封装协议，是一种面向比特位的数据链路层协议，如图 6-1 所示。Cisco 默认的广域网封装协议便是 HDLC，不过 Cisco 在原有 HDLC 的基础上进行优化，实现多协议环境的支持。但是当思科跟其他厂商的设备进行广域网连接时，没法很好实现兼容，此时需要采用其他广域网封装协议如 PPP 协议。

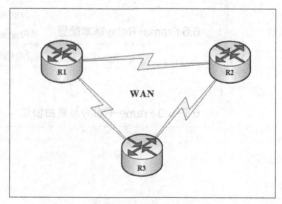

图 6-1　广域网连接

【HDLC】
高级链路数据控制协议是广域网中最简单的协议，实现基本的数据联通。

【DCE& DTE】
在广域网同步串行通信中，通信双方需要有相同的时钟速率进行通信。一般电信局端为 DCE 端，即数据信息终端，用户接入端为 DTE 端，即数据终端设备。一般 DCE 端提供时钟速率。在 GNS3 环境下，无需为设备部署时钟速率。

6.1　HDLC 基本配置

实验步骤：

1. 依据图中拓扑，为 R1 和 R2 的串口链路部署 HDLC 封装和 IP 地址。

R1 上配置如下所示。

```
R1(config)#int s0/0
// 进入端口配置模式，S 是 "serial" 的缩写，即路由器中的串行接口。
Serial 口用来传输的数据是同步的，连接时候需要 DCE 同步时钟速率
R1(config-if)#no shutdown
// 开启端口
R1(config-if)#encapsulation hdlc
// 串口默认采用 HDLC 封装
R1(config-if)#ip address 12.1.1.1 255.255.255.0
// 配置 IP 地址
R1(config-if)#exit
```

R2 上配置如下所示。

```
R2(config)#int s0/0
R2(config-if)#no shutdown
R2(config-if)#encapsulation hdlc
R2(config-if)#ip address 12.1.1.2 255.255.255.0
R2(config-if)#exit
```

2. 查看接口封装情况，并测试直连连通性，如下所示。

```
R1#show interfaces s0/0
// 查看串口的物理链路信息

Serial0/0 is up, line protocol is up
// 端口的物理层和协议层都起来了
  Hardware is M4T
  Internet address is 12.1.1.1/24
  MTU 1500 bytes, BW 1544 Kbit, DLY 20000 usec,
     reliability 255/255, txload 1/255, rxload 1/255
  Encapsulation HDLC, crc 16, loopback not set
// High-Level Data Link Control (HDLC) is a bit-oriented code-transparent
synchronous data link layer protocol
  Keepalive set (10 sec)
  Restart-Delay is 0 secs
  Last input 00:00:01, output 00:00:09, output hang never
  Last clearing of "show interface" counters never
```

```
Input queue: 0/75/0/0 (size/max/drops/flushes); Total output drops: 0
Queueing strategy: weighted fair
Output queue: 0/1000/64/0 (size/max total/threshold/drops)
   Conversations  0/1/256 (active/max active/max total)
   Reserved Conversations 0/0 (allocated/max allocated)
   Available Bandwidth 1158 kilobits/sec
5 minute input rate 0 bits/sec, 0 packets/sec
5 minute output rate 0 bits/sec, 0 packets/sec
   24 packets input, 2262 bytes, 0 no buffer
   Received 24 broadcasts, 0 runts, 0 giants, 0 throttles
   0 input errors, 0 CRC, 0 frame, 0 overrun, 0 ignored, 0 abort
   31 packets output, 2982 bytes, 0 underruns
   0 output errors, 0 collisions, 3 interface resets
   0 output buffer failures, 0 output buffers swapped out
   4 carrier transitions   DCD=up DSR=up DTR=up RTS=up CTS=up
```

可以看到，路由器采用 HDLC 协议进行封装，并且接口处于正常状态，在 R1 上 Ping R2，如下所示。

```
R1#ping 12.1.1.2
Type escape sequence to abort.
Sending 5, 100-byte ICMP Echos to 12.1.1.2, timeout is 2 seconds:
!!!!!

// 连接成功
Success rate is 100 percent (5/5), round-trip min/avg/max = 20/25/40 ms
```

此时，通过 HDLC 封装后，直连连通性正常。

3. 查看 HDLC 封装格式，如图 6 - 2 所示。

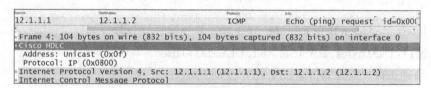

图 6-2　HDLC 分组

可以看到，此处的二层封装协议是 Cisco HDLC，而不是官方的 HDLC。Cisco 路由器默认采用便是私有的 HDLC 封装，与其他厂商不兼容，若广域网链路有其他厂商的设备，则需要采用 PPP 协议进行封装。此实验完成。

6.2 PPP 基本配置

实验目的：
1. 掌握 PPP 基本配置。
2. 理解 PPP 封装格式。

实验拓扑：

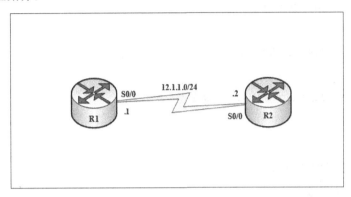

实验随手记：

实验原理：

1. PPP 概述

PPP（Point-to-Point Protocol，点对点协议）是广域网点对点串行链路上的封装协议。相比 HDLC 协议，PPP 协议具备更好的拓展性功能，例如链路捆绑、身份认证、数据压缩、错误校验、多协议拓展等。

2. PPP 组成

PPP 由多个子协议组成，如 LCP 和 NCP 子协议。LCP（link control protocol，链路控制协议）用于管理链路连通性，NCP<network control protocol，网络控制协议> 用于实现网络层支持，如图 6-3 所示。

图 6-3 PPP 协议

【PPP】
PPP 协议是广域网协议中使用最广泛的协议之一，它具备非常好的拓展功能。

实验步骤：

1. 依据图中拓扑，为 R1 和 R2 的串口链路部署 PPP 封装和 IP 地址。

R1 上配置如下所示。

R1(config)#int s0/0

// 进入端口配置模式，S 是 "serial" 的缩写，即路由器中的串行接口。
Serial 口用来传输的数据是同步的，连接时候需要 DCE 同步时钟频率

R1(config-if)#no shutdown

// 开启端口

R1(config-if)#encapsulation ppp

// 封装 PPP 协议，Point-to-Point Protocol，工作在数据链路层

R1(config-if)#ip address 12.1.1.1 255.255.255.0

// 配置 IP 地址

R1(config-if)#exit

R2 上配置如下所示。

R2(config)#int s0/0

R2(config-if)#no shutdown

R2(config-if)#encapsulation ppp

R2(config-if)#ip address 12.1.1.2 255.255.255.0

R2(config-if)#exit

---

2. 查看接口封装情况，并测试直连连通性，如下所示。

R1#show interfaces s0/0

// 查看 S0/0 这个串口的信息

Serial0/0 is up, line protocol is up

// 端口的物理层和协议层都起来了

  Hardware is M4T

  Internet address is 12.1.1.1/24

  MTU 1500 bytes, BW 1544 Kbit, DLY 20000 usec,

    reliability 255/255, txload 1/255, rxload 1/255

  Encapsulation PPP, LCP Open

// 成功地封装了 PPP 协议

  Open: IPCP, CDPCP, crc 16, loopback not set

  Keepalive set (10 sec)

  Restart-Delay is 0 secs

  Last input 00:00:25, output 00:00:08, output hang never

  Last clearing of "show interface" counters 00:01:09

  Input queue: 0/75/0/0 (size/max/drops/flushes); Total output drops: 0

```
Queueing strategy: weighted fair
Output queue: 0/1000/64/0 (size/max total/threshold/drops)
   Conversations  0/1/256 (active/max active/max total)
   Reserved Conversations 0/0 (allocated/max allocated)
   Available Bandwidth 1158 kilobits/sec
5 minute input rate 0 bits/sec, 0 packets/sec
5 minute output rate 0 bits/sec, 0 packets/sec
   25 packets input, 1394 bytes, 0 no buffer
   Received 0 broadcasts, 0 runts, 0 giants, 0 throttles
   0 input errors, 0 CRC, 0 frame, 0 overrun, 0 ignored, 0 abort
   26 packets output, 1699 bytes, 0 underruns
   0 output errors, 0 collisions, 1 interface resets
   0 output buffer failures, 0 output buffers swapped out
   1 carrier transitions    DCD=up DSR=up DTR=up RTS=up CTS=up
```

可以看到，路由器采用 PPP 协议进行封装，并且接口处于正常状态，在 R1 上 Ping R2，如下所示。

```
R1#ping 12.1.1.2
// 测试连通性
Type escape sequence to abort.
Sending 5, 100-byte ICMP Echos to 12.1.1.2, timeout is 2 seconds:
!!!!!
// 测试成功！
Success rate is 100 percent (5/5), round-trip min/avg/max = 20/25/40 ms
```

此时，通过 PPP 封装后，直连连通性正常。

3. 查看 PPP 封装格式，如图 6 - 4 所示。

```
Source        Destination     Protocol    Info
12.1.1.1      12.1.1.2        ICMP        Echo (ping) request  id=0x00
12.1.1.2      12.1.1.1        ICMP        Echo (ping) reply    id=0x00
Frame 5: 104 bytes on wire (832 bits), 104 bytes captured (832 bits) on interface 0
Point-to-Point Protocol
   Address: 0xff
   Control: 0x03
   Protocol: Internet Protocol version 4 (0x0021)
Internet Protocol Version 4, Src: 12.1.1.1 (12.1.1.1), Dst: 12.1.1.2 (12.1.1.2)
Internet Control Message Protocol
```

图 6-4 PPP 分组

相比 HDLC 协议，PPP 协议支持更多的功能，例如多协议支持、身份认证、链路捆绑等，并且兼容不同厂商，是更为优秀的广域网封装协议。此实验完成。

6.3 PPP PAP 认证

实验目的：
1. 掌握 PPP PAP 认证基本配置。
2. 理解 PAP 的单向认证和双向认证。
3. 理解 PAP 的明文加密方式。

实验拓扑：

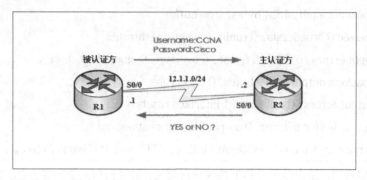

实验随手记：

_____
_____
_____
_____
_____
_____

【明文认证】
认证过程中密码直接以未加密的方式出现在链路中，容易被黑客直接截取密码并攻击。

实验原理：
PPP 可以支持 PAP 和 CHAP 两种安全认证方法，两种认证方式有以下特征。

PPP 认证	PAP 认证	CHAP 认证
加密类型	明文认证	密文认证
认证方式	单向或双向认证	双向认证
握手方式	两次握手	三次握手

由于 PAP 认证采取明文认证，认证密钥可以通过数据截取获得，非常不安全，所以强烈不建议采用 PAP 进行安全认证。

【密文认证】
认证过程中密码通过加密方式进行交互，防止被黑客窃取，更加安全。

实验步骤：
1. 依据图中拓扑，为 R1 和 R2 部署 PPP 封装和 IP 地址。
R1 上配置如下所示。

R1(config)#int s0/0
// 进入端口配置模式，S 是 "serial" 的缩写，即路由器中的串行接口

Serial 口用来传输的数据是同步的，连接时需要 DCE 同步时钟频率

R1(config-if)#no shutdown

// 开启端口

R1(config-if)#encapsulation ppp

// 封装 PPP 协议，Point-to-Point Protocol，工作在数据链路层。

R1(config-if)#ip address 12.1.1.1 255.255.255.0

// 配置 IP 地址

R1(config-if)#exit

R2 上配置如下所示。

R2(config)#int s0/0

R2(config-if)#no shutdown

R2(config-if)#encapsulation ppp

R2(config-if)#ip address 12.1.1.2 255.255.255.0

R2(config-if)#exit

测试连通性，如下所示。

R1#ping 12.1.1.2

Type escape sequence to abort.

Sending 5, 100-byte ICMP Echos to 12.1.1.2, timeout is 2 seconds:

!!!!!

// 测试连通性，成功

Success rate is 100 percent (5/5), round-trip min/avg/max = 20/26/44 ms

2. 部署 PAP 单向认证，其中 R2 为主认证方，R1 为被认证方。

R2 上配置如下所示。

R2(config)#username CCNA password Cisco

// 定义本地用户名数据库，用于实现安全认证

R2(config)#int s0/0

R2(config-if)#ppp authentication pap

// 开启接口下 PAP 认证

R1 上配置如下所示。

R1(config)#int s0/0

R1(config-if)#ppp pap sent-username CCNA password Cisco

//PPP 被认证方需要将用户名密码发送给认证方

单向认证中，一边为认证方，一边为被认证方。

3. 部署 PAP 双向认证，在上面配置的基础上加入以下配置。

R1 上配置如下所示。

R1(config)#username CCNA2 password Cisco2
// 把 R1 作为主认证方
R1(config)#int s0/0
R1(config-if)#ppp authentication pap
// 开启接口下 PAP 认证

R2 上配置如下所示。

R2(config)#int s0/0
R2(config-if)#ppp pap sent-username CCNA2 password Cisco2
//PPP 被认证方需要将用户名密码发送给认证方

双向认证中，R1 和 R2 同为认证方和被认证方。

通过抓包理解 PAP 认证方式，如图 6-5 所示。

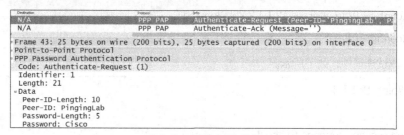

图 6-5　PPP PAP 分组

从 PPP 分组视图可以看出，PAP 认证采用明文方式，用户名和密码可以看到，所以非常不安全！在当今网络安全问题越来越严峻的情况下，非常不推荐采用此认证方式。此实验完成。

6.4 PPP CHAP 认证

实验目的：
1. 掌握 PPP CHAP 认证基本配置。
2. 理解 CHAP 的双向认证。
3. 理解 CHAP 的密文认证方式。

实验拓扑：

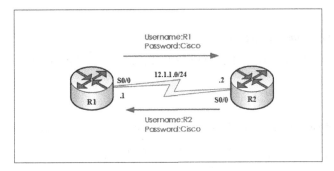

实验随手记：

_____
_____
_____
_____
_____
_____

实验步骤：

1. 依据图中拓扑，为 R1 和 R2 部署 PPP 封装和 IP 地址。

R1 上配置如下所示。

```
R1(config)#int s0/0
// 进入端口配置模式，S 是 "serial" 的缩写，即路由器中的串行接口
Serial 口用来传输的数据是同步的，连接时需要 DCE 同步时钟频率
R1(config-if)#no shutdown
// 开启端口
R1(config-if)#encapsulation ppp
// 封装 PPP 协议，Point-to-Point Protocol，工作在数据链路层。
R1(config-if)#ip address 12.1.1.1 255.255.255.0
// 配置 IP 地址
R1(config-if)#exit
```

R2 上配置如下所示。

```
R2(config)#int s0/0
R2(config-if)#no shutdown
R2(config-if)#encapsulation ppp
R2(config-if)#ip address 12.1.1.2 255.255.255.0
R2(config-if)#exit
```

测试连通性,如下所示。

```
R1#ping 12.1.1.2
// 连通性测试
Type escape sequence to abort.
Sending 5, 100-byte ICMP Echos to 12.1.1.2, timeout is 2 seconds:
!!!!!
// 成功
Success rate is 100 percent (5/5), round-trip min/avg/max = 20/26/44 ms
```

2. 部署 CHAP 认证,实现安全认证,配置如下所示。

```
R1(config)#username R2 password Cisco
// 添加本地账号密码
R1(config)#int s0/0
R1(config-if)#ppp authentication chap
// 开启 PPP 的 CHAP 认证
R1(config-if)#exit
R2(config)#username R1 password Cisco
// 同 R1
R2(config)#int s0/0
R2(config-if)#ppp authentication chap
R2(config-if)#exit
```

CHAP 认证采用双向认证,用户名是对方的主机名,密码相同。

3. 抓取 CHAP 认证分组,理解 CHAP 密文认证方式,如图 6-6 所示。

```
Destination                    Protocol    Info
N/A                            PPP CHAP    Challenge (NAME='R1', VALUE=0xb9027492491614b
N/A                            PPP CHAP    Response (NAME='R2', VALUE=0xa82cb35ad754322c
N/A                            PPP CHAP    Success (MESSAGE='')
▶ Frame 9: 27 bytes on wire (216 bits), 27 bytes captured (216 bits) on interface 0
▶ Point-to-Point Protocol
▼ PPP Challenge Handshake Authentication Protocol
    Code: Challenge (1)
    Identifier: 1
    Length: 23
  ▼ Data
      Value Size: 16
      Value: b9027492491614becda522d472de328b
      Name: R1
```

图 6-6　PPP CHAP 分组

可以看到,CHAP 认证过程无法看到认证密码,可以保证认证的安全性。相比 PAP 的明文认证,强烈推荐 CHAP 认证。此实验完成。

6.5　PPP Multilink

实验目的：
1. 掌握 PPP 多链路捆绑的基本配置。
2. 理解 PPP 多链路捆绑的功能。

实验拓扑：

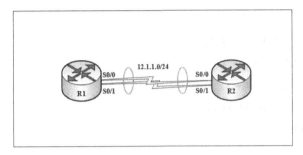

实验随手记：

_____

_____

_____

_____

_____

实验原理：

PPP Multilink< 多链路捆绑 > 技术可以将多根广域网串口链路进行捆绑，提高链路带宽，增强网络冗余。PPP Multilink 技术与 Etherchannel（以太通道）非常类似，只不过前者用于广域网，后者用于局域网。

实验步骤：

1. 依据图中拓扑，为 R1 和 R2 的各个串口部署 PPP 封装。

R1 上配置如下所示。

```
R1(config)#int s0/0
// 进入端口配置模式，S 是 "serial" 的缩写，即路由器中的串行接口
Serial 口用来传输的数据是同步的，连接时需要 DCE 同步时钟频率
R1(config-if)#no shutdown
// 开启端口
R1(config-if)#encapsulation ppp
// 封装 PPP 协议，Point-to-Point Protocol，工作在数据链路层
R1(config-if)#int s0/1
```

// 这里有个小细节，就是没有退出就进入到另外一个端口了，这样的操作是被允许的
R1(config-if)#no shutdown
R1(config-if)#encapsulation ppp
R1(config-if)#exit

R2 上配置如下所示。

R2(config)#int s0/0
R2(config-if)#no shutdown
R2(config-if)#encapsulation ppp
R2(config-if)#int s0/1
R2(config-if)#no shutdown
R2(config-if)#encapsulation ppp
R2(config-if)#exit
// 同 R1

2. 部署 PPP 捆绑。

R1 上配置如下所示。

R1(config)#int s0/0
R1(config-if)#ppp multilink
// 开启 PPP 多链路捆绑
R1(config-if)#ppp multilink group 1
// 将接口放入捆绑组，这里的 group 用于区分本地多个捆绑组，是本地标识
R1(config-if)#int s0/1
// 同样的操作，也把 s0/1 端口放入 group 1 中
R1(config-if)#ppp multilink
R1(config-if)#ppp multilink group 1
R1(config-if)#exit

R2 上配置如下所示。

R2(config)#int s0/0
R2(config-if)#ppp multilink
R2(config-if)#ppp multilink group 1
R2(config-if)#int s0/1
R2(config-if)#ppp multilink
R2(config-if)#ppp multilink group 1
R2(config-if)#exit

3. 为逻辑捆绑口配置 IP 地址，如下所示。

R1(config)#int multilink 1
// 进入逻辑口，为其分配 IP 地址
R1(config-if)#ip add 12.1.1.1 255.255.255.0

```
// 由于是逻辑端口，所以默认是打开的，不需要再次"no shutdown"打开，与
loopback 口一样
R1(config-if)#exit


R2(config)#int multilink 1
R2(config-if)#ip add 12.1.1.2 255.255.255.0
R2(config-if)#exit
```

4. 测试 PPP 多链路捆绑。

① 查看接口 IP 状态，如下所示。

```
R1#show interfaces multilink 1
// 查看逻辑端口 1 的主要信息
Multilink1 is up, line protocol is up
  Hardware is multilink group interface
  Internet address is 12.1.1.1/24
  MTU 1500 bytes, BW 3088 Kbit, DLY 100000 usec,
//Serial 口为 1544Kbit，逻辑口刚好是其 2 倍
     reliability 255/255, txload 1/255, rxload 1/255
  Encapsulation PPP, LCP Open, multilink Open
  Open: IPCP, CDPCP, loopback not set
  Keepalive set (10 sec)
  DTR is pulsed for 2 seconds on reset
  Last input 00:00:10, output never, output hang never
  Last clearing of "show interface" counters 00:10:04
  Input queue: 0/75/0/0 (size/max/drops/flushes); Total output drops: 0
  Queueing strategy: fifo
  Output queue: 0/40 (size/max)
  5 minute input rate 0 bits/sec, 0 packets/sec
  5 minute output rate 0 bits/sec, 0 packets/sec
     32 packets input, 5126 bytes, 0 no buffer
     Received 0 broadcasts, 0 runts, 0 giants, 0 throttles
     0 input errors, 0 CRC, 0 frame, 0 overrun, 0 ignored, 0 abort
     32 packets output, 5496 bytes, 0 underruns
     0 output errors, 0 collisions, 2 interface resets
     0 output buffer failures, 0 output buffers swapped out
     0 carrier transitions
```

可以看到，逻辑捆绑链路的带宽是原来 Serial 的 2 倍。

② 查看接口 IP 信息，如下所示。

```
R1#show ip int brief
Interface         IP-Address      OK? Method Status                Protocol
Serial0/0         unassigned      YES unset  up                    up
Serial0/1         unassigned      YES unset  up                    up
Serial0/2         unassigned      YES unset  administratively down down
Serial0/3         unassigned      YES unset  administratively down down
Multilink1        12.1.1.1        YES manual up                    up
```

③测试连通性，如下所示。

```
R1#ping 12.1.1.2
// 测试连通性
Type escape sequence to abort.
Sending 5, 100-byte ICMP Echos to 12.1.1.2, timeout is 2 seconds:
!!!!!
// 成功
Success rate is 100 percent (5/5), round-trip min/avg/max = 16/33/48 ms
```

测试成功，说明通过部署 PPP 多链路捆绑，可以实现链路冗余和带宽聚合。此实验完成。

6.6 Frame-Relay 基本配置

实验目的：
1. 掌握帧中继的基本配置。
2. 掌握通过路由器模拟帧中继交换机。
3. 理解帧中继的映射表、转发表等。

实验拓扑：

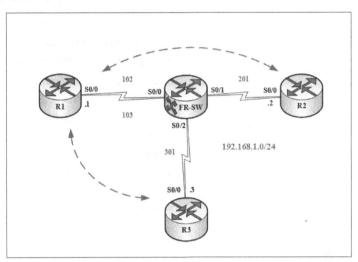

实验随手记：

_____
_____
_____
_____
_____

实验原理：

1. Frame-Relay 概述

帧中继技术是一种基于分组交换的广域网技术，它是从 X.25 协议改进而来的。相比专线服务，帧中继可以实现性价比更高的服务，它的拓展性非常高，可以支持星型拓扑、部分互联拓扑、全互联拓扑等各种网络连接需求。于此同时，帧中继技术采用虚电路 PVC 将不同客户进行逻辑隔离，保证数据的安全性。

2. Frame-Relay 术语

术语	解释
PVC	Permanent Virtual Circuit，永久虚电路指的是运营商在帧中继网络中为客户指定的虚拟通道

【Frame-Relay】
帧中继的前身是 X.25 技术，都是基于二层的标签或标识号进行数据转发。

续表

术语	解释
DLCI	Data Link Connection Identifier，数据链路连接标识符。DLCI 是 FR 技术的二层地址，类似 Ethernet 技术的 MAC 地址，范围从 0 ~ 1023，0 ~ 15 保留
LMI	Local Manager Interface，本地管理接口客户端设备与运营商设备之间的链路管理机制，实现链路连通性检测等功能
IARP	Inverse ARP，逆向 ARP，用于实现对方 IP 与本地 DLCI 的映射，类似 ARP 映射表
DTE	Data Terminal Equipment，数据终端设备，一般指客户设备端，如客户网络边缘路由器
DCE	Data Communication Equipment，数据通信设备，一般指运营商设备端，如帧中继交换机。DCE 向 DTE 提供时钟速率
帧中继转发表	运营商设备 DCE 端，即帧中继交换机的数据转发表，类似以太网交换机的 CAM 表，保存 DLCI 到接口的映射
帧中继映射表	客户端 DTE 设备保存的表项，保存 DLCI 到 IP 的映射

【帧中继交换机】帧中继交换机是真帧中继环境中用于实现数据转发的设备。它具备帧中继转发表，存储着 DLCI 与接口的映射信息，类似以太网交换机中的 CAM 表。

实验步骤：

1. 依据图中拓扑，通过路由器模拟帧中继交换机。

①开启帧中继交换功能，如下所示。

FW-SW(config)#frame-relay switching

// 若没有开启帧中继交换功能，后续转发表无法正常工作

②接口开启帧中继封装，并定义为 DCE 接口，如下所示。

FW-SW(config)#int s0/0

// 进入 s0/0 的端口

FW-SW(config-if)#no shutdown

FW-SW(config-if)#encapsulation frame-relay

// 接口封装帧中继协议

FW-SW(config-if)#frame-relay intf-type dce

// 接口类型设置为 DCE 接口，一般运营商端为 DCE，本地局端为 DTE

FW-SW(config-if)#exit

FW-SW(config)#int s0/1

FW-SW(config-if)#no shutdown

FW-SW(config-if)#encapsulation frame-relay

FW-SW(config-if)#frame-relay intf-type dce

```
FW-SW(config-if)#exit
FW-SW(config)#int s0/2
FW-SW(config-if)#no shutdown
FW-SW(config-if)#encapsulation frame-relay
FW-SW(config-if)#frame-relay intf-type dce
FW-SW(config-if)#exit
```

③编写帧中继转发条目，如下所示。

```
FW-SW(config)#int s0/0
FW-SW(config-if)#frame-relay route 102 interface s0/1 201
// 手工编写帧中继转发表，将进接口的进标签转换为出接口的出标签，帧中继交换机
会对二层数据帧进行修改，不同于以太网交换机
FW-SW(config-if)#frame-relay route 103 interface s0/2 301
FW-SW(config-if)#exit
FW-SW(config)#int s0/1
FW-SW(config-if)#frame-relay route 201 interface s0/0 102
FW-SW(config-if)#exit
FW-SW(config)#int s0/2
FW-SW(config-if)#frame-relay route 301 interface s0/0 103
FW-SW(config-if)#exit
```

2. 通过部署帧中继技术，各个站点可以互相连通，其中 R1 为中心点，R2 和 R3 为分支点。

R1 上配置如下所示。

```
R1(config)#int s0/0
R1(config-if)#no shutdown
R1(config-if)#encapsulation frame-relay
// 把端口封装成帧中继
R1(config-if)#no frame-relay inverse-arp
// 关闭 IARP，即逆向 ARP
R1(config-if)#frame-relay map ip 192.168.1.2 102 broadcast
// 一般做映射时都要加入 Broadcast 参数，用于支持帧中继环境下运行动态路由协议，
broadcast 表示此接口能向外发送组播或者广播分组
R1(config-if)#frame-relay map ip 192.168.1.3 103 broadcast
R1(config-if)#exit
// 逆向 ARP 采用周期性方式工作，每 60s 交互一次，不同于 ARP 的触发式工作方式，
所以通信会有延迟。不仅如此，不同设备对 IARP 的支持不同，若 IARP 出现问题，映
射失败，则数据封装失败，对通信有很大影响，所以，一般关闭 IARP 并且手工编写
DLCI 到 IP 的映射，更加稳妥！
```

R2 上配置如下所示。

R2(config)#int s0/0

R2(config-if)#no shutdown

R2(config-if)#encapsulation frame-relay

R2(config-if)#no frame-relay inverse-arp

R2(config-if)#frame-relay map ip 192.168.1.1 201 broadcast

R2(config-if)#exit

R3 上配置如下所示。

R3(config)#int s0/0

R3(config-if)#no shutdown

R3(config-if)#encapsulation frame-relay

R3(config-if)#no frame-relay inverse-arp

R3(config-if)#frame-relay map ip 192.168.1.1 301 broadcast

R3(config-if)#exit

3. 查看帧中继的运行状态。

①查看 PVC 状态，如下所示。

R1#show frame-relay pvc

// 查看帧中继端口的状况

PVC Statistics for interface Serial0/0 (Frame Relay DTE)

	Active	Inactive	Deleted	Static
Local	2	0	0	0
Switched	0	0	0	0
Unused	0	0	0	0

DLCI = 102, DLCI USAGE = LOCAL, PVC STATUS = ACTIVE, INTERFACE = Serial0/0

//PVC 状态有三种，不同状态标识不同情形，具体如下。

1.ACTIVE：表示 PVC 完全正常。

2.INACITVE：表示本端 PVC 正常，对端不正常。

3.DELETED：表示本端 PVC 故障。

input pkts 89	output pkts 47	in bytes 8364
out bytes 6032	dropped pkts 0	in pkts dropped 0
out pkts dropped 0	out bytes dropped 0	
in FECN pkts 0	in BECN pkts 0	out FECN pkts 0
out BECN pkts 0	in DE pkts 0	out DE pkts 0
out bcast pkts 47	out bcast bytes 6032	

```
5 minute input rate 0 bits/sec, 0 packets/sec
5 minute output rate 0 bits/sec, 0 packets/sec
pvc create time 00:45:16, last time pvc status changed 00:18:37

DLCI = 103, DLCI USAGE = LOCAL, PVC STATUS = ACTIVE, INTERFACE = Serial0/0

  input pkts 0           output pkts 10          in bytes 0
  out bytes 1100         dropped pkts 0          in pkts dropped 0
  out pkts dropped 0     out bytes dropped 0
  in FECN pkts 0         in BECN pkts 0          out FECN pkts 0
  out BECN pkts 0        in DE pkts 0            out DE pkts 0
  out bcast pkts 10      out bcast bytes 1100
  5 minute input rate 0 bits/sec, 0 packets/sec
  5 minute output rate 0 bits/sec, 0 packets/sec
  pvc create time 00:45:18, last time pvc status changed 00:01:58
```

PVC 状态处于 Active 状态，说明 PVC 连接没有问题。

② 查看帧中继的映射信息，如下所示。

```
R1#show frame-relay map
// 查看帧中继的映射信息
Serial0/0 (up): ip 192.168.1.2 dlci 102(0x66,0x1860), static,
//Static 说明是静态映射
          broadcast,
          CISCO, status defined, active
Serial0/0 (up): ip 192.168.1.3 dlci 103(0x67,0x1870), static,
          broadcast,
          CISCO, status defined, active
```

可以看到，由于关闭 IARP 并且手工绑定 DLCI 与 IP 的映射，所以这里显示 Static，并且处于 Active 状态，说明映射没有问题。

③ 查看帧中继交换机上的转发表，如下所示。

```
FW-SW#show frame-relay route
// 查看帧中继的路由信息
Input Intf    Input Dlci    Output Intf    Output Dlci    Status
Serial0/0     102           Serial0/1      201            active
Serial0/0     103           Serial0/2      301            active
Serial0/1     201           Serial0/0      102            active
Serial0/2     301           Serial0/0      102            active
```

4. 测试连通性，如下所示。

```
R1#ping 192.168.1.2

Type escape sequence to abort.
Sending 5, 100-byte ICMP Echos to 192.168.1.2, timeout is 2 seconds:
!!!!!
// 成功
Success rate is 100 percent (5/5), round-trip min/avg/max = 36/42/60 ms
R1#ping 192.168.1.3

Type escape sequence to abort.
Sending 5, 100-byte ICMP Echos to 192.168.1.3, timeout is 2 seconds:
!!!!!
Success rate is 100 percent (5/5), round-trip min/avg/max = 16/31/52 ms
```

以上说明总部和分支之间的连通性没有问题，接着测试分支之间的连通性，如下所示。

```
R2#ping 192.168.1.3

Type escape sequence to abort.
Sending 5, 100-byte ICMP Echos to 192.168.1.3, timeout is 2 seconds:
.....
Success rate is 0 percent (0/5)
```

可以看到，分支之间没法相互通信，查看 R2 和 R3 的映射表，如下所示。

```
R2#show frame-relay map
Serial0/0 (up): ip 192.168.1.1 dlci 201(0xC9,0x3090), static,
                broadcast,
                CISCO, status defined, active
R3#show frame-relay map
Serial0/0 (up): ip 192.168.1.1 dlci 301(0x12D,0x48D0), static,
                broadcast,
                CISCO, status defined, active
```

由于分支之间没有到对方的映射信息，数据没法正常封装，需要加入分支对方的映射，如下所示。

```
R2(config)#int s0/0
R2(config-if)#frame-relay map ip 192.168.1.3 201 broadcast
R2(config-if)#exit
R3(config)#int s0/0
```

```
R3(config-if)#frame-relay map ip 192.168.1.2 301 broadcast
R3(config-if)#exit
```

再次查看映射表，如下所示。

```
R2#show frame-relay map
Serial0/0 (up): ip 192.168.1.1 dlci 201(0xC9,0x3090), static,
              broadcast,
              CISCO, status defined, active
Serial0/0 (up): ip 192.168.1.3 dlci 201(0xC9,0x3090), static,
              broadcast,
              CISCO, status defined, active
R3#show frame-relay map
Serial0/0 (up): ip 192.168.1.1 dlci 301(0x12D,0x48D0), static,
              broadcast,
              CISCO, status defined, active
Serial0/0 (up): ip 192.168.1.2 dlci 301(0x12D,0x48D0), static,
              broadcast,
              CISCO, status defined, active
```

继续测试分支之间的连通性，如下所示。

```
R2#ping 192.168.1.3

Type escape sequence to abort.
Sending 5, 100-byte ICMP Echos to 192.168.1.3, timeout is 2 seconds:
!!!!!
Success rate is 100 percent (5/5), round-trip min/avg/max = 44/68/96 ms
```

当加入映射后，分支正常通信！注意，星形拓扑中分支之间的数据需要经过中心点进行转发。此实验完成。

6.7　Frame-Relay&Static Route

实验目的：
1. 掌握帧中继上部署静态路由技术。
2. 通过静态路由实现不同分支之间的通信。

实验拓扑：

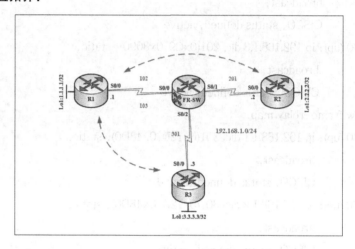

实验随手记：

实验原理：
　　通过在帧中继环境下部署静态路由，可以实现分支之间的相互通信。静态路由的部署相对来说比较简单，可以采用出接口和下一跳的方式进行部署。但是在帧中继环境下，只能采用下一跳 IP 地址的方式来编写，否则通信会出现问题。帧中继环境下，通信不仅仅要考虑三层路由，还需要考虑二层映射问题。当采用出接口方式进行编写的时候，三层路由有了，但是出接口无法递归到具体的 DLCI 号，此时二层封装失败。例如，以【ip route 1.1.1.1 255.255.255.255 S0/0】来编写，路由器知道往 1.1.1.1 走时，需要往接口 S0/0 发送数据，重点是：采用哪个 DLCI 号呢？但如果以【ip route 1.1.1.1 255.255.255.255 192.168.1.1】来编写，路由器知道往 1.1.1.1 走时，需要将数据丢给 192.168.1.1，再去找 192.168.1.1 对应的 DLCI 号。若路由器本地映射表存在【192.168.1.1 - 201】的映射（一般直连网段的地址通过 IARP 或者手工方式已经完成映射），此时可完成封装，数据便可以向外发送。

6.7 Frame-Relay&Static Route

实验步骤：

1. 依据图中拓扑，通过路由器模拟帧中继交换机。

①开启帧中继交换功能，如下所示。

```
FW-SW(config)#frame-relay switching
// 若没有开启帧中继交换功能，后续转发表无法正常工作
```

②接口开启帧中继封装，并定义为 DCE 接口，如下所示。

```
FW-SW(config)#int s0/0
FW-SW(config-if)#no shutdown
FW-SW(config-if)#encapsulation frame-relay
// 接口封装帧中继协议
FW-SW(config-if)#frame-relay intf-type dce
// 接口类型设置为 DCE 接口，一般运营商端为 DCE，本地局端为 DTE
FW-SW(config-if)#exit
FW-SW(config)#int s0/1
FW-SW(config-if)#no shutdown
FW-SW(config-if)#encapsulation frame-relay
FW-SW(config-if)#frame-relay intf-type dce
FW-SW(config-if)#exit
FW-SW(config)#int s0/2
FW-SW(config-if)#no shutdown
FW-SW(config-if)#encapsulation frame-relay
FW-SW(config-if)#frame-relay intf-type dce
FW-SW(config-if)#exit
```

③编写帧中继转发条目，如下所示。

```
FW-SW(config)#int s0/0
FW-SW(config-if)#frame-relay route 102 interface s0/1 201
// 手工编写帧中继转发表，将进接口的进标签转换为出接口的出标签，帧中继交换机
会对二层数据帧进行修改，不同于以太网交换机
FW-SW(config-if)#frame-relay route 103 interface s0/2 301
FW-SW(config-if)#exit
FW-SW(config)#int s0/1
FW-SW(config-if)#frame-relay route 201 interface s0/0 102
FW-SW(config-if)#exit
FW-SW(config)#int s0/2
FW-SW(config-if)#frame-relay route 301 interface s0/0 103
FW-SW(config-if)#exit
```

2. 通过部署帧中继技术，使得各个站点直连连通，其中 R1 为中心点，R2 和 R3 为分支点。

R1 上配置如下所示。

```
R1(config)#int s0/0
R1(config-if)#no shutdown
R1(config-if)#encapsulation frame-relay
R1(config-if)#no frame-relay inverse-arp
// 关闭 IARP，即逆向 ARP
R1(config-if)#frame-relay map ip 192.168.1.2 102 broadcast
// 一般做映射时都要加入 Broadcast 参数，用于支持帧中继环境下运行动态路由协议，
broadcast 表示此接口能向外发送组播或者广播分组
R1(config-if)#frame-relay map ip 192.168.1.3 103 broadcast
R1(config-if)#exit
```

R2 上配置如下所示。

```
R2(config)#int s0/0
R2(config-if)#no shutdown
R2(config-if)#encapsulation frame-relay
R2(config-if)#no frame-relay inverse-arp
R2(config-if)#frame-relay map ip 192.168.1.1 201 broadcast
R2(config-if)#frame-relay map ip 192.168.1.3 201 broadcast
R2(config-if)#exit
```

R3 上配置如下所示。

```
R3(config)#int s0/0
R3(config-if)#no shutdown
R3(config-if)#encapsulation frame-relay
R3(config-if)#no frame-relay inverse-arp
R3(config-if)#frame-relay map ip 192.168.1.1 301 broadcast
R3(config-if)#frame-relay map ip 192.168.1.2 301 broadcast
R3(config-if)#exit
```

测试直连连通性，如下所示。

```
R1#ping 192.168.1.2

Type escape sequence to abort.
Sending 5, 100-byte ICMP Echos to 192.168.1.2, timeout is 2 seconds:
!!!!!
Success rate is 100 percent (5/5), round-trip min/avg/max = 36/42/60 ms
R1#ping 192.168.1.3
```

```
Type escape sequence to abort.
Sending 5, 100-byte ICMP Echos to 192.168.1.3, timeout is 2 seconds:
!!!!!
Success rate is 100 percent (5/5), round-trip min/avg/max = 16/31/52 ms
R2#ping 192.168.1.3


Type escape sequence to abort.
Sending 5, 100-byte ICMP Echos to 192.168.1.3, timeout is 2 seconds:
!!!!!
Success rate is 100 percent (5/5), round-trip min/avg/max = 44/68/96 ms
```

可以看到，直连连通没有问题。

3. 部署静态路由技术，使得不同分支之间能够相互通信，配置如下所示。

```
R1(config)#ip route 2.2.2.2 255.255.255.255 192.168.1.2
// 帧中继环境下，静态路由写法需要采用下一跳。如果采用出接口写法，则无法找到映
射信息，封装失败，导致通信失败。若采用下一跳，则经过路由递归后可以找到映射信息。
一般在点对点环境下静态路由采用出接口写法，在多路访问环境下采用下一跳写法
R1(config)#ip route 3.3.3.3 255.255.255.255 192.168.1.3

R2(config)#ip route 1.1.1.1 255.255.255.255 192.168.1.1
R2(config)#ip route 3.3.3.3 255.255.255.255 192.168.1.3

R3(config)#ip route 1.1.1.1 255.255.255.255 192.168.1.1
R3(config)#ip route 2.2.2.2 255.255.255.255 192.168.1.2
```

此时测试不同分支背后网段能否相互通信，如下所示。

```
R1#ping 2.2.2.2 source 1.1.1.1


Type escape sequence to abort.
Sending 5, 100-byte ICMP Echos to 2.2.2.2, timeout is 2 seconds:
Packet sent with a source address of 1.1.1.1
!!!!!
// 成功
Success rate is 100 percent (5/5), round-trip min/avg/max = 24/45/64 ms
R1#ping 3.3.3.3 source 1.1.1.1


Type escape sequence to abort.
Sending 5, 100-byte ICMP Echos to 3.3.3.3, timeout is 2 seconds:
Packet sent with a source address of 1.1.1.1
```

```
!!!!!
Success rate is 100 percent (5/5), round-trip min/avg/max = 40/42/52 ms
R2#ping 3.3.3.3 source 2.2.2.2
Type escape sequence to abort.
Sending 5, 100-byte ICMP Echos to 3.3.3.3, timeout is 2 seconds:
Packet sent with a source address of 2.2.2.2
!!!!!
Success rate is 100 percent (5/5), round-trip min/avg/max = 48/59/80 ms
```

测试成功，此实验完成。

6.8 Frame-Relay&RIPv2

实验目的：

1. 掌握帧中继上部署 RIPv2。
2. 通过部署 RIPv2 实现不同分支之间的通信。

实验拓扑：

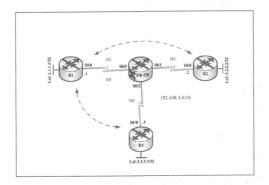

实验随手记：

_____
_____
_____
_____
_____
_____

实验原理：

1.帧中继网络是一种 NBMA（非广播多路访问）网络，不支持组播或广播数据，而动态路由协议一般采用组播或广播进行通信。为了实现在帧中继下运行动态路由协议，一般在路由器上执行映射的时候，加入"broadcast"参数，用于支持伪广播，使得路由协议运行起来。

2.距离矢量协议如 RIP 和 EIGRP 在帧中继环境下会因为水平分割特性出现通信问题，例如分支之间的路由条目无法相互学到。可以通过两种方法来解决此问题：

①关闭水平分割。
②划分子接口。
由于 RIP 在帧中继环境下默认开启水平分割特性，所以不需要做特殊处理。

实验步骤：

1.依据图中拓扑，通过路由器模拟帧中继交换机。
①开启帧中继交换功能，如下所示。

```
FW-SW(config)#frame-relay switching
```
// 若没有开启帧中继交换功能，后续转发表无法正常工作

②接口开启帧中继封装，并定义为 DCE 接口，如下所示。

```
FW-SW(config)#int s0/0
FW-SW(config-if)#no shutdown
FW-SW(config-if)#encapsulation frame-relay
```
// 接口封装帧中继协议
```
FW-SW(config-if)#frame-relay intf-type dce
```
// 接口类型设置为 DCE 接口，一般运营商端为 DCE，本地局端为 DTE
```
FW-SW(config-if)#exit
FW-SW(config)#int s0/1
FW-SW(config-if)#no shutdown
FW-SW(config-if)#encapsulation frame-relay
FW-SW(config-if)#frame-relay intf-type dce
FW-SW(config-if)#exit
FW-SW(config)#int s0/2
FW-SW(config-if)#no shutdown
FW-SW(config-if)#encapsulation frame-relay
FW-SW(config-if)#frame-relay intf-type dce
FW-SW(config-if)#exit
```

③编写帧中继转发条目，如下所示。

```
FW-SW(config)#int s0/0
FW-SW(config-if)#frame-relay route 102 interface s0/1 201
```
// 手工编写帧中继转发表，将进接口的进标签转换为出接口的出标签，帧中继交换机会对二层数据帧进行修改，不同于以太网交换机
```
FW-SW(config-if)#frame-relay route 103 interface s0/2 301
FW-SW(config-if)#exit
FW-SW(config)#int s0/1
FW-SW(config-if)#frame-relay route 201 interface s0/0 102
FW-SW(config-if)#exit
FW-SW(config)#int s0/2
FW-SW(config-if)#frame-relay route 301 interface s0/0 103
FW-SW(config-if)#exit
```

2. 通过部署帧中继技术，各个站点直连连通，其中 R1 为中心点，R2 和 R3 为分支点。

R1 上配置如下所示。

```
R1(config)#int s0/0
R1(config-if)#no shutdown
R1(config-if)#encapsulation frame-relay
// 把端口封装成帧中继
R1(config-if)#no frame-relay inverse-arp
// 关闭 IARP, 即逆向 ARP
R1(config-if)#frame-relay map ip 192.168.1.2 102 broadcast
// 一般做映射时都要加入 Broadcast 参数, 用于支持帧中继环境下运行动态路由协议,
broadcast 表示此接口能向外发送组播或者广播分组
R1(config-if)#frame-relay map ip 192.168.1.3 103 broadcast
R1(config-if)#exit
```

R2 上配置如下所示。

```
R2(config)#int s0/0
R2(config-if)#no shutdown
R2(config-if)#encapsulation frame-relay
R2(config-if)#no frame-relay inverse-arp
R2(config-if)#frame-relay map ip 192.168.1.1 201 broadcast
R2(config-if)#frame-relay map ip 192.168.1.3 201 broadcast
R2(config-if)#exit
```

R3 上配置如下所示。

```
R3(config)#int s0/0
R3(config-if)#no shutdown
R3(config-if)#encapsulation frame-relay
R3(config-if)#no frame-relay inverse-arp
R3(config-if)#frame-relay map ip 192.168.1.1 301 broadcast
R3(config-if)#frame-relay map ip 192.168.1.2 301 broadcast
R3(config-if)#exit
```

测试直连连通性，如下所示。

```
R1#ping 192.168.1.2

Type escape sequence to abort.
Sending 5, 100-byte ICMP Echos to 192.168.1.2, timeout is 2 seconds:
!!!!!
// 成功!
Success rate is 100 percent (5/5), round-trip min/avg/max = 36/42/60 ms
R1#ping 192.168.1.3
```

```
Type escape sequence to abort.
Sending 5, 100-byte ICMP Echos to 192.168.1.3, timeout is 2 seconds:
!!!!!
Success rate is 100 percent (5/5), round-trip min/avg/max = 16/31/52 ms
R2#ping 192.168.1.3

Type escape sequence to abort.
Sending 5, 100-byte ICMP Echos to 192.168.1.3, timeout is 2 seconds:
!!!!!
Success rate is 100 percent (5/5), round-trip min/avg/max = 44/68/96 ms
```

可以看到，直连连通没有问题。

3. 部署 RIPv2 路由协议，使得不同分支之间能够相互通信，配置如下所示。

```
R1(config)#router rip
// 开启 rip 协议进程
R1(config-router)#version 2
// 设置成 RIPv2 版本
R1(config-router)#no auto-summary
// 关闭自动汇总功能
R1(config-router)#network 1.0.0.0
// 宣告网段，无需带子网掩码
R1(config-router)#network 192.168.1.0
R1(config-router)#exit

R2(config)#router rip
R2(config-router)#version 2
R2(config-router)#no auto-summary
R2(config-router)#network 2.0.0.0
R2(config-router)#network 192.168.1.0
R2(config-router)#exit

R3(config)#router rip
R3(config-router)#version 2
R3(config-router)#no auto-summary
R3(config-router)#network 3.0.0.0
R3(config-router)#network 192.168.1.0
R3(config-router)#exit
```

查看路由表，如下所示。

R1#show ip route rip

// 查看有关 RIP 协议的路由信息

    2.0.0.0/32 is subnetted, 1 subnets

R   2.2.2.2 [120/1] via 192.168.1.2, 00:00:01, Serial0/0

    3.0.0.0/32 is subnetted, 1 subnets

R   3.3.3.3 [120/1] via 192.168.1.3, 00:00:17, Serial0/0

R2#show ip route rip

    1.0.0.0/32 is subnetted, 1 subnets

R   1.1.1.1 [120/1] via 192.168.1.1, 00:00:15, Serial0/0

    3.0.0.0/32 is subnetted, 1 subnets

R   3.3.3.3 [120/2] via 192.168.1.3, 00:00:15, Serial0/0

R3#show ip route rip

    1.0.0.0/32 is subnetted, 1 subnets

R   1.1.1.1 [120/1] via 192.168.1.1, 00:00:18, Serial0/0

    2.0.0.0/32 is subnetted, 1 subnets

R   2.2.2.2 [120/2] via 192.168.1.2, 00:00:18, Serial0/0

可以看到，全局路由信息已经同步。测试连通性，如下所示。

R1#ping 2.2.2.2 source 1.1.1.1

Type escape sequence to abort.

Sending 5, 100-byte ICMP Echos to 2.2.2.2, timeout is 2 seconds:

Packet sent with a source address of 1.1.1.1

!!!!!

// 成功

Success rate is 100 percent (5/5), round-trip min/avg/max = 16/52/80 ms

R1#ping 3.3.3.3 source 1.1.1.1

Type escape sequence to abort.

Sending 5, 100-byte ICMP Echos to 3.3.3.3, timeout is 2 seconds:

Packet sent with a source address of 1.1.1.1

!!!!!

Success rate is 100 percent (5/5), round-trip min/avg/max = 16/25/36 ms

R2#ping 3.3.3.3 source 2.2.2.2

```
Type escape sequence to abort.
Sending 5, 100-byte ICMP Echos to 3.3.3.3, timeout is 2 seconds:
Packet sent with a source address of 2.2.2.2
!!!!!
Success rate is 100 percent (5/5), round-trip min/avg/max = 24/52/84 ms
```

测试成功，说明在帧中继上部署 RIPv2 可以实现不同分支互联。但是需要提醒一点，在默认情况下，运行 RIPv2 时接口下关闭水平分割，如下所示。

```
R1#show ip interface s0/0
Serial0/0 is up, line protocol is up
  Internet address is 192.168.1.1/24
  Broadcast address is 255.255.255.255
  Address determined by non-volatile memory
  MTU is 1500 bytes
  Helper address is not set
  Directed broadcast forwarding is disabled
  Multicast reserved groups joined: 224.0.0.9
  Outgoing access list is not set
  Inbound  access list is not set
  Proxy ARP is enabled
  Local Proxy ARP is disabled
  Security level is default
  Split horizon is disabled
// 关闭水平分割
……
```

若水平分割开启，则分支之间的路由信息学习会出现问题，在之后的实验中有解答。此实验完成。

6.9 Frame-Relay&EIGRP

实验目的：
1. 掌握帧中继上部署 EIGRP。
2. 通过部署 EIGRP 实现不同分支之间的通信。
3. 理解帧中继子接口，包括点对点和多点子接口。

实验拓扑：

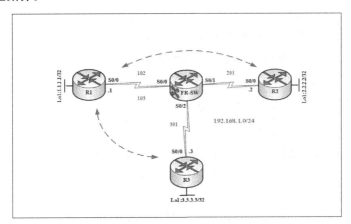

实验随手记：

_____
_____
_____
_____
_____

实验步骤：

1. 依据图中拓扑，通过路由器模拟帧中继交换机。

①开启帧中继交换功能，如下所示。

FW-SW(config)#frame-relay switching
// 若没有开启帧中继交换功能，后续转发表无法正常工作

②接口开启帧中继封装，并定义为 DCE 接口，如下所示。

FW-SW(config)#int s0/0
FW-SW(config-if)#no shutdown
FW-SW(config-if)#encapsulation frame-relay
// 接口封装帧中继协议
FW-SW(config-if)#frame-relay intf-type dce
// 接口类型设置为 DCE 接口，一般运营商端为 DCE，本地局端为 DTE

```
FW-SW(config-if)#frame-relay intf-type dce
FW-SW(config-if)#exit
FW-SW(config)#int s0/1
FW-SW(config-if)#no shutdown
FW-SW(config-if)#encapsulation frame-relay
FW-SW(config-if)#frame-relay intf-type dce
FW-SW(config-if)#exit
FW-SW(config)#int s0/2
FW-SW(config-if)#no shutdown
FW-SW(config-if)#encapsulation frame-relay
FW-SW(config-if)#frame-relay intf-type dce
FW-SW(config-if)#exit
```

③编写帧中继转发条目，如下所示。

```
FW-SW(config)#int s0/0
FW-SW(config-if)#frame-relay route 102 interface s0/1 201
// 手工编写帧中继转发表，将进接口的进标签转换为出接口的出标签，帧中继交换机
会对二层数据帧进行修改，不同于以太网交换机
FW-SW(config-if)#frame-relay route 103 interface s0/0 301
FW-SW(config-if)#exit
FW-SW(config)#int s0/1
FW-SW(config-if)#frame-relay route 201 interface s0/0 102
FW-SW(config-if)#exit
FW-SW(config)#int s0/2
FW-SW(config-if)#frame-relay route 301 interface s0/0 103
FW-SW(config-if)#exit
```

2. 通过部署帧中继技术，各个站点直连连通，其中 R1 为中心点，R2 和 R3 为分支点。

R1 上配置如下所示。

```
R1(config)#int s0/0
R1(config-if)#no shutdown
R1(config-if)#encapsulation frame-relay
// 把端口封装成帧中继
R1(config-if)#no frame-relay inverse-arp
// 关闭 IARP，即逆向 ARP
R1(config-if)#frame-relay map ip 192.168.1.2 102 broadcast
```

// 一般做映射时都要加入 Broadcast 参数，用于支持帧中继环境下运行动态路由协议。broadcast 表示此接口能向外发送组播或者广播分组。

R1(config-if)#frame-relay map ip 192.168.1.3 103 broadcast
R1(config-if)#exit

　　R2 上配置如下所示。

R2(config)#int s0/0
R2(config-if)#no shutdown
R2(config-if)#encapsulation frame-relay
R2(config-if)#no frame-relay inverse-arp
R2(config-if)#frame-relay map ip 192.168.1.1 201 broadcast
R2(config-if)#frame-relay map ip 192.168.1.3 201 broadcast
R2(config-if)#exit

　　R3 上配置如下所示。

R3(config)#int s0/0
R3(config-if)#no shutdown
R3(config-if)#encapsulation frame-relay
R3(config-if)#no frame-relay inverse-arp
R3(config-if)#frame-relay map ip 192.168.1.1 301 broadcast
R3(config-if)#frame-relay map ip 192.168.1.2 301 broadcast
R3(config-if)#exit

　　测试直连连通性，如下所示。

R1#ping 192.168.1.2

Type escape sequence to abort.
Sending 5, 100-byte ICMP Echos to 192.168.1.2, timeout is 2 seconds:
!!!!!
// 成功
Success rate is 100 percent (5/5), round-trip min/avg/max = 36/42/60 ms
R1#ping 192.168.1.3

Type escape sequence to abort.
Sending 5, 100-byte ICMP Echos to 192.168.1.3, timeout is 2 seconds:
!!!!!
Success rate is 100 percent (5/5), round-trip min/avg/max = 16/31/52 ms
R2#ping 192.168.1.3

```
Type escape sequence to abort.
Sending 5, 100-byte ICMP Echos to 192.168.1.3, timeout is 2 seconds:
!!!!!
Success rate is 100 percent (5/5), round-trip min/avg/max = 44/68/96 ms
```

可以看到，直连连通没有问题。

3. 部署 EIGRP 路由协议。

R1 上配置如下所示。

```
R1(config)#router eigrp 100
// 开启 EIGRP 协议进程
R1(config-router)#no auto-summary
// 关闭自动汇总
R1(config-router)#network 192.168.1.0
// 宣告网段
R1(config-router)#network 1.0.0.0
R1(config-router)#exit
```

R2 上配置如下所示。

```
R2(config)#router eigrp 100
R2(config-router)#no auto-summary
R2(config-router)#network 192.168.1.0
R2(config-router)#network 2.0.0.0
R2(config-router)#exit
```

R3 上配置如下所示。

```
R3(config)#router eigrp 100
R3(config-router)#no auto-summary
R3(config-router)#network 192.168.1.0
R3(config-router)#network 3.0.0.0
R3(config-router)#exit
```

此时查看路由表。

R1 上显示如下所示。

```
R1#show ip route eigrp
     2.0.0.0/32 is subnetted, 1 subnets
D       2.2.2.2 [90/2297856] via 192.168.1.2, 00:02:40, Serial0/0
     3.0.0.0/32 is subnetted, 1 subnets
D       3.3.3.3 [90/2297856] via 192.168.1.3, 00:00:26, Serial0/0
```

R2 上显示如下所示。

```
R2#show ip route eigrp
    1.0.0.0/32 is subnetted, 1 subnets
D   1.1.1.1 [90/2297856] via 192.168.1.1, 00:02:47, Serial0/0
```

R3 上显示如下所示。

```
R3#show ip route eigrp
    1.0.0.0/32 is subnetted, 1 subnets
D   1.1.1.1 [90/2297856] via 192.168.1.1, 00:02:43, Serial0/0
```

可以看到，中心点 R1 可以学到 R2 和 R3 的路由条目，但是分支点 R2 和 R3 之间没有相互学到对方路由，这是由接口的水平分割特性所造成的："从本接口学到的路由条目不会从本接口发送出去。"为了解决由水平分割所导致的路由信息不同步问题，可以有两种解决方案：①接口下关闭水平分割；②创建逻辑子接口。

4. 关闭接口水平分割特性，使得分支之间相互学习到路由，配置如下所示。

```
R1(config)#int s0/0
R1(config-if)#no ip split-horizon eigrp 100
// 接口下关闭水平分割时必须加入 EIGRP 进程，否则无效
```

查看 R2 和 R3 上的路由表，如下所示。

```
R2#show ip route eigrp
    1.0.0.0/32 is subnetted, 1 subnets
D   1.1.1.1 [90/2297856] via 192.168.1.1, 00:15:12, Serial0/0
    3.0.0.0/32 is subnetted, 1 subnets
D   3.3.3.3 [90/2809856] via 192.168.1.1, 00:00:36, Serial0/0
// 学到了路由
R3#show ip route eigrp
    1.0.0.0/32 is subnetted, 1 subnets
D   1.1.1.1 [90/2297856] via 192.168.1.1, 00:15:06, Serial0/0
    2.0.0.0/32 is subnetted, 1 subnets
D   2.2.2.2 [90/2809856] via 192.168.1.1, 00:00:39, Serial0/0
```

此时，分支之间相互学习到路由。

5. 创建逻辑子接口，将不同的 PVC 映射到子接口，其中 R1 和 R2 处在 192.168.1.0/24 网段，R1 和 R3 处在 192.168.2.0/24 网段。

R1 上配置如下所示。

```
R1(config)#default int s0/0
// 初始化接口配置
R1(config)#int s0/0
R1(config-if)#encapsulation frame-relay
R1(config)#int s0/0.1 point-to-point
// 帧中继子接口分两种：
```

1. 点对点子接口；
2. 多点子接口。

其中点对点子接口主要用于点对点拓扑环境，而多点子接口与帧中继主接口类似，可以用于多路访问环境。

R1(config-subif)#ip address 192.168.1.1 255.255.255.0

R1(config-subif)#frame-relay interface-dlci 102

// 点对点子接口的映射不需要对方 IP 地址，只需要映射本端 DLCI 即可

R1(config-fr-dlci)#exit

R1(config)#int s0/0.2 point-to-point

R1(config-subif)#ip add 192.168.2.1 255.255.255.0

R1(config-subif)#frame-relay interface-dlci 103

R1(config-fr-dlci)#exit

R1(config)#router eigrp 100

R1(config-router)#network 192.168.2.0

R1(config-router)#exit

R3 上配置如下所示。

R3(config)#int s0/0

R3(config-if)#ip address 192.168.2.3 255.255.255.0

R3(config-if)#frame-relay map ip 192.168.2.1 301 broadcast

// 重新映射

R3(config)#router eigrp 100

R3(config-router)#network 192.168.2.0

R3(config-router)#exit

再次查看 R2 和 R3 的路由表，如下所示。

R2#show ip route eigrp

    1.0.0.0/32 is subnetted, 1 subnets

D   1.1.1.1 [90/2297856] via 192.168.1.1, 00:08:10, Serial0/0

    3.0.0.0/32 is subnetted, 1 subnets

D   3.3.3.3 [90/2809856] via 192.168.1.1, 00:05:10, Serial0/0

D   192.168.2.0/24 [90/2681856] via 192.168.1.1, 00:05:13, Serial0/0

R3#show ip route eigrp

    1.0.0.0/32 is subnetted, 1 subnets

D   1.1.1.1 [90/2297856] via 192.168.2.1, 00:05:15, Serial0/0

    2.0.0.0/32 is subnetted, 1 subnets

D   2.2.2.2 [90/2809856] via 192.168.2.1, 00:05:15, Serial0/0

D   192.168.1.0/24 [90/2681856] via 192.168.2.1, 00:05:15, Serial0/0

可以看到，通过部署逻辑子接口，分支之间可以相互学习到对方路由。此实验完成。

第 7 章　安全策略

本章主要学习安全策略技术，包括访问控制列表和地址转换技术。其中，访问控制列表包含了标准 ACL、拓展 ACL、时间 ACL、自反 ACL；而地址转换技术包含了动态 NAT、端口 NAT 和静态 NAT。ACL 和 NAT 两种技术在企业网络和校园网络中的应用非常广泛，是最基础的安全策略。以下是本章导航图：

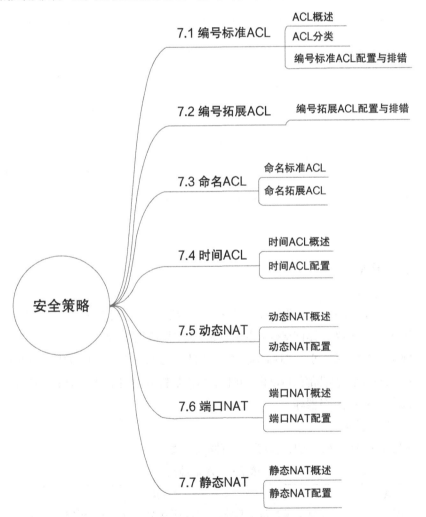

7.1 编号标准 ACL

实验目的：

1. 掌握编号标准 ACL 的基本编写。
2. 理解编号标准 ACL 的基本特性。

实验拓扑：

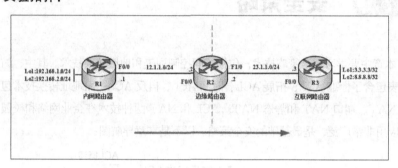

实验随手记：

_____

_____

_____

_____

_____

实验原理：

1. ACL 概述

ACL（Access-list，访问控制列表）是一种流量安全过滤工具，可以抓取特定流量并执行丢弃和转发，是网络技术中众多策略表中最基础的一种。ACL 的使用非常广泛，例如，通过 ACL 可以实现局域网中不同部门之间的访问限制，如技术部和财务部不能相互访问，其他部门可以相互访问；可以限制某个部门不能上网；可以限制特定主机访问特定的服务器等。

2. ACL 分类

根据特征的不同，ACL 可以分为 4 种类型，如表 7-1 所示。

表 7-1 ACL 种类

ACL	功能
编号 ACL	ACL 编写基于编号，无法删除编号内特定某条语句
命名 ACL	ACL 编写基于命名，可以修改、删除、增加特定某条语句
标准 ACL	ACL 只能匹配源 IP 地址
拓展 ACL	ACL 能匹配源目 IP 地址、源目 Port、协议等

【ACL】
访问控制列表是网络工程师接触的第一种列表，用于实现流量过滤。类似功能有分发列表、过滤列表等技术。

根据 ACL 的特征，可以这么理解，命名是编号 ACL 的升级，拓展是标准 ACL 的升级。在实际使用过程中，一般两两特性组合成一种 ACL，比如：编号标准 ACL、编号拓展 ACL、命名标准 ACL、命名拓展 ACL 等。

实验步骤：

1. 依据图中拓扑，配置各个路由器的 IP 地址，并部署静态路由保证全网连通。

R1 上配置如下所示。

```
R1(config)#ip route 23.1.1.0 255.255.255.0 12.1.1.2
R1(config)#ip route 3.3.3.3 255.255.255.255 12.1.1.2
R1(config)#ip route 8.8.8.8 255.255.255.255 12.1.1.2
```

R2 上配置如下所示。

```
R2(config)#ip route 192.168.1.0 255.255.255.0 12.1.1.1
R2(config)#ip route 192.168.2.0 255.255.255.0 12.1.1.1
R2(config)#ip route 3.3.3.3 255.255.255.255 23.1.1.3
R2(config)#ip route 8.8.8.8 255.255.255.255 23.1.1.3
```

R3 上配置如下所示。

```
R3(config)#ip route 12.1.1.0 255.255.255.0 23.1.1.2
R3(config)#ip route 192.168.1.0 255.255.255.0 23.1.1.2
R3(config)#ip route 192.168.2.0 255.255.255.0 23.1.1.2
```

测试连通性，如下所示。

```
R1#ping 3.3.3.3 source 192.168.1.1

Type escape sequence to abort.
Sending 5, 100-byte ICMP Echos to 3.3.3.3, timeout is 2 seconds:
Packet sent with a source address of 192.168.1.1
!!!!!
// 成功
Success rate is 100 percent (5/5), round-trip min/avg/max = 40/40/40 ms
R1#ping 8.8.8.8 source 192.168.2.1

Type escape sequence to abort.
Sending 5, 100-byte ICMP Echos to 8.8.8.8, timeout is 2 seconds:
Packet sent with a source address of 192.168.2.1
!!!!!
Success rate is 100 percent (5/5), round-trip min/avg/max = 36/43/56 ms
```

可以看到，内网与外网通信没有问题。

2. 在 R2 上部署编号标准 ACL，使得内网网段 192.168.1.0/24 不能访问外网，其他都能访问，配置如下所示。

```
R2(config)#access-list 1 deny 192.168.1.0 0.0.0.255
// 标准 ACL 只能匹配源，不能匹配目的地。标准 ACL 的编号范围是 1~99 和 1300~1999
// 其中"1"是该条访问控制列表的编号。Deny 表示阻止从哪里发过来数据。后面接的是反掩码
R2(config)#access-list 1 permit any
// 由于一旦使用访问列表之后，就会默认 deny 掉所有流量，因而需要在最后加上一句 permit any，表示放行其他的流量
R2(config)#int f0/0
R2(config-if)#ip access-group 1 in
// 接口的 in 和 out 方向要以本地路由器为参考点，然后考虑流量方向
R2(config-if)#exit
```

测试 ACL 的效果，让内网路由器 R1 访问外网路由器 R3，如下所示。

```
R1#ping 3.3.3.3 source 192.168.1.1

Type escape sequence to abort.
Sending 5, 100-byte ICMP Echos to 3.3.3.3, timeout is 2 seconds:
Packet sent with a source address of 192.168.1.1
UUUUU
//U 表示数据不可达或者被禁止，Unreachable
Success rate is 0 percent (0/5)
R1#ping 8.8.8.8 source 192.168.1.1

Type escape sequence to abort.
Sending 5, 100-byte ICMP Echos to 8.8.8.8, timeout is 2 seconds:
Packet sent with a source address of 192.168.1.1
UUUUU
Success rate is 0 percent (0/5)
R1#ping 3.3.3.3 source 192.168.2.1

Type escape sequence to abort.
Sending 5, 100-byte ICMP Echos to 3.3.3.3, timeout is 2 seconds:
Packet sent with a source address of 192.168.2.1
!!!!!
Success rate is 100 percent (5/5), round-trip min/avg/max = 16/38/64 ms
R1#ping 8.8.8.8 source 192.168.2.1
```

```
Type escape sequence to abort.
Sending 5, 100-byte ICMP Echos to 8.8.8.8, timeout is 2 seconds:
Packet sent with a source address of 192.168.2.1
!!!!!
Success rate is 100 percent (5/5), round-trip min/avg/max = 20/38/48 ms
```

从上面可以很明显地看出，内网网段 192.168.1.0/24 无法访问外网，其他网段可以。

3. 在 R2 上部署编号标准 ACL，使得 IP 地址 192.168.2.1 可以 Telnet 本地，而其他 IP 地址没法访问，实现安全管理。

R2 上配置如下所示。

```
R2(config)#line vty 0 15
R2(config-line)#no login
// 无需验证，主要方便实验测试，就是不需要密码的意思
R2(config)#access-list 2 permit host 192.168.2.1
R2(config)#line vty 0 15
R2(config-line)#access-class 2 in
//VTY 下调用 ACL，一般是 IN 方向
```

此时在 R1 和 R3 上 Telnet R2，如下所示。

```
R1#telnet 12.1.1.2
Trying 12.1.1.2 ...
% Connection refused by remote host
// 表示连接被拒绝
R3#telnet 23.1.1.2
Trying 23.1.1.2 ...
% Connection refused by remote host
R1#telnet 12.1.1.2 /source-interface lo2
Trying 12.1.1.2 ... Open
R2>
```

可以看到，只有 192.168.2.1 才能远程访问 Telnet。编号标准 ACL 可以实现最基础的访问限制和安全管理，但是在限制要求比较高的情况下则无能为力。若上面的要求"192.168.1.0 无法访问互联网"修改成"192.168.1.0 无法访问 3.3.3.3，可以访问 8.8.8.8"，则编号标准 ACL 无法实现，因为标准 ACL 只能匹配源，不能匹配目的以及具体的端口和协议。此实验完成。

7.2 编号拓展 ACL

实验目的：
1. 掌握编号拓展 ACL 的基本编写。
2. 理解编号拓展 ACL 的基本特性。

实验拓扑：

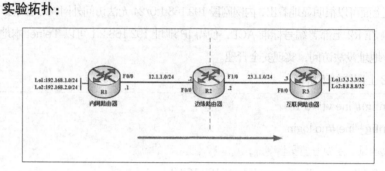

实验随手记：

实验步骤：

1. 依据图中拓扑，配置各个路由器的 IP 地址，并部署静态路由保证全网连通。
R1 上配置如下所示。

```
R1(config)#ip route 23.1.1.0 255.255.255.0 12.1.1.2
R1(config)#ip route 3.3.3.3 255.255.255.255 12.1.1.2
R1(config)#ip route 8.8.8.8 255.255.255.255 12.1.1.2
```

R2 上配置如下所示。

```
R2(config)#ip route 192.168.1.0 255.255.255.0 12.1.1.1
R2(config)#ip route 192.168.2.0 255.255.255.0 12.1.1.1
R2(config)#ip route 3.3.3.3 255.255.255.255 23.1.1.3
R2(config)#ip route 8.8.8.8 255.255.255.255 23.1.1.3
```

R3 上配置如下所示。

```
R3(config)#ip route 12.1.1.0 255.255.255.0 23.1.1.2
R3(config)#ip route 192.168.1.0 255.255.255.0 23.1.1.2
```

R3(config)#ip route 192.168.2.0 255.255.255.0 23.1.1.2

测试连通性，如下所示。

R1#ping 3.3.3.3 source 192.168.1.1

Type escape sequence to abort.
Sending 5, 100-byte ICMP Echos to 3.3.3.3, timeout is 2 seconds:
Packet sent with a source address of 192.168.1.1
!!!!!
Success rate is 100 percent (5/5), round-trip min/avg/max = 40/40/40 ms
R1#ping 8.8.8.8 source 192.168.2.1

Type escape sequence to abort.
Sending 5, 100-byte ICMP Echos to 8.8.8.8, timeout is 2 seconds:
Packet sent with a source address of 192.168.2.1
!!!!!
Success rate is 100 percent (5/5), round-trip min/avg/max = 36/43/56 ms

可以看到，内网与外网通信没有问题。

2. 在 R2 上部署编号拓展 ACL，使得内网网段 192.168.1.0/24 不能访问外网主机 3.3.3.3 的 23 号端口，192.168.2.0/24 网段不能 Ping 通 8.8.8.8。

R2 上配置如下所示。

R2(config)#access-list 100 deny tcp 192.168.1.0 0.0.0.255 host 3.3.3.3 eq 23
// 拓展控制列表，可以匹配或拒绝源和目的 IP、源和目的端口等，可以实现更高要求的访问控制。

R2(config)#access-list 100 deny icmp 192.168.2.0 0.0.0.255 host 8.8.8.8
// 拒绝 192.168.2.0/24 网段访问主机 8.8.8.8

R2(config)#access-list 100 permit ip any any
// 同标准的访问列表一样，也需要在最后放行其他的流量通过

R2(config)#int f0/0
R2(config-if)#ip access-group 100 in
// 把访问列表应用在这个端口上

R3 上配置如下所示。

R3(config)#line vty 0 15
R3(config-line)#no login
R3(config-line)#exit

3. 测试编号拓展 ACL，如下所示。

R1#telnet 3.3.3.3 /source-interface lo1

// 注意，这里表示 R1 采用源地址为本地环回 IP 来访问 3.3.3.3

Trying 3.3.3.3 ...

% Destination unreachable; gateway or host down

// 目标不可达，网关或者主机挂掉了。

R1#telnet 3.3.3.3

// 当采用本地物理接口访问 3.3.3.3 则直接进入

Trying 3.3.3.3 ... Open

R3>exit

可以看到，R1 本地环回地址 192.168.1.1 无法远程访问 3.3.3.3，其他地址则可以。

R1#ping 8.8.8.8

Type escape sequence to abort.

Sending 5, 100-byte ICMP Echos to 8.8.8.8, timeout is 2 seconds:

!!!!!

// 成功！

Success rate is 100 percent (5/5), round-trip min/avg/max = 36/40/44 ms

R1#ping 8.8.8.8 source loopback 2

Type escape sequence to abort.

Sending 5, 100-byte ICMP Echos to 8.8.8.8, timeout is 2 seconds:

Packet sent with a source address of 192.168.2.1

UUUUU

// 被阻断了！

Success rate is 0 percent (0/5)

可以看出，192.168.2.0 网段无法 Ping 通 8.8.8.8，但是其他网段可以。

从上面的实验可以分析得到，编号拓展 ACL 可以匹配源目 IP、端口和协议。相比标准 ACL 来说，对流量的控制更加细腻，能满足更加复杂的网络环境。但是只要是基于编号的 ACL，不管是编号标准还是编号拓展，都不易于管理，例如无法删除 ACL 中单独的某条语句等，而基于命名的 ACL 可以解决这些问题，接下来有详细介绍。此实验完成。

7.3 命名 ACL

实验目的:
1. 掌握命名 ACL 的基本编写。
2. 理解命名 ACL 的基本特性。
3. 理解命名 ACL 和编号 ACL 的区别。

实验拓扑:

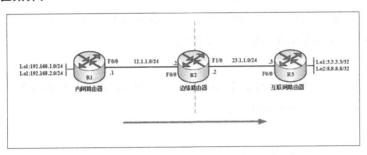

实验随手记:

_____
_____
_____
_____
_____
_____

实验步骤:

1. 依据图中拓扑，配置各个路由器的 IP 地址，并部署静态路由保证全网连通。
R1 上配置如下所示。

```
R1(config)#ip route 23.1.1.0 255.255.255.0 12.1.1.2
R1(config)#ip route 3.3.3.3 255.255.255.255 12.1.1.2
R1(config)#ip route 8.8.8.8 255.255.255.255 12.1.1.2
```

R2 上配置如下所示。

```
R2(config)#ip route 192.168.1.0 255.255.255.0 12.1.1.1
R2(config)#ip route 192.168.2.0 255.255.255.0 12.1.1.1
R2(config)#ip route 3.3.3.3 255.255.255.255 23.1.1.3
R2(config)#ip route 8.8.8.8 255.255.255.255 23.1.1.3
```

R3 上配置如下所示。

```
R3(config)#ip route 12.1.1.0 255.255.255.0 23.1.1.2
R3(config)#ip route 192.168.1.0 255.255.255.0 23.1.1.2
R3(config)#ip route 192.168.2.0 255.255.255.0 23.1.1.2
```

测试连通性，如下所示。

```
R1#ping 3.3.3.3 source 192.168.1.1

Type escape sequence to abort.
Sending 5, 100-byte ICMP Echos to 3.3.3.3, timeout is 2 seconds:
Packet sent with a source address of 192.168.1.1
!!!!!
Success rate is 100 percent (5/5), round-trip min/avg/max = 40/40/40 ms
R1#ping 8.8.8.8 source 192.168.2.1

Type escape sequence to abort.
Sending 5, 100-byte ICMP Echos to 8.8.8.8, timeout is 2 seconds:
Packet sent with a source address of 192.168.2.1
!!!!!
Success rate is 100 percent (5/5), round-trip min/avg/max = 36/43/56 ms
```

可以看到，内网与外网通信没有问题。

2. 在 R2 上部署基于命名的标准 ACL，使得内网网段 192.168.1.0/24 不能访问外网，其他都能访问，配置如下所示。

```
R2(config)#ip access-list standard DENYVLAN10
// standard 表示这个是标准访问列表，DENYVLAN10 是这个访问列表的名字
R2(config-std-nacl)#deny 192.168.1.0 0.0.0.255
R2(config-std-nacl)#permit any
R2(config-std-nacl)#exit
R2(config)#int f0/0
R2(config-if)#ip access-group DENYVLAN10 in
// 把它应用在 f0/0 这个端口上，方向是进入路由的方向
R2(config-if)#exit
```

测试 ACL 的效果，让内网路由器 R1 访问外网路由器 R3，如下所示。

```
R1#ping 3.3.3.3 source 192.168.1.1
// 带源地址 PING
Type escape sequence to abort.
Sending 5, 100-byte ICMP Echos to 3.3.3.3, timeout is 2 seconds:
Packet sent with a source address of 192.168.1.1
U.U.U
```

```
// 被阻断了!
Success rate is 0 percent (0/5)
R1#ping 8.8.8.8 source 192.168.1.1

Type escape sequence to abort.
Sending 5, 100-byte ICMP Echos to 8.8.8.8, timeout is 2 seconds:
Packet sent with a source address of 192.168.1.1
UUUUU
Success rate is 0 percent (0/5)
R1#ping 3.3.3.3 source 192.168.2.1

Type escape sequence to abort.
Sending 5, 100-byte ICMP Echos to 3.3.3.3, timeout is 2 seconds:
Packet sent with a source address of 192.168.2.1
!!!!!
// 成功了
Success rate is 100 percent (5/5), round-trip min/avg/max = 16/38/64 ms
R1#ping 8.8.8.8 source 192.168.2.1

Type escape sequence to abort.
Sending 5, 100-byte ICMP Echos to 8.8.8.8, timeout is 2 seconds:
Packet sent with a source address of 192.168.2.1
!!!!!
Success rate is 100 percent (5/5), round-trip min/avg/max = 20/38/48 ms
```

从上面可以明显地看出，内网网段 192.168.1.0/24 无法访问外网，其他网段可以。

3. 在 R2 上部署命名拓展 ACL，使得内网网段 192.168.1.0/24 不能访问外网主机 3.3.3.3 的 23 号端口，192.168.2.0/24 网段不能 Ping 通 8.8.8.8。

R2 上配置如下所示。

```
R2(config)#ip access-list extended DENYSERVICE
//extended 表示拓展 ACL，DENYSERVICE 是名字，建议用大写，以免和命令弄混了
R2(config-ext-nacl)#deny tcp 192.168.1.0 0.0.0.255 host 3.3.3.3 eq 23
R2(config-ext-nacl)#deny icmp 192.168.2.0 0.0.0.255 host 8.8.8.8
R2(config-ext-nacl)#permit ip any any
R2(config-ext-nacl)#exit
R2(config)#int f0/0
R2(config-if)#ip access-group DENYSERVICE in
// 把 DNEYSERVICE 这个拓展 ACL 部署在 f0/0 这个端口中
```

R3 上配置如下所示。

R3(config)#line vty 0 15

R3(config-line)#no login

R3(config-line)#exit

测试命名拓展 ACL，如下所示。

R1#telnet 3.3.3.3 /source-interface lo1

Trying 3.3.3.3 ...

% Destination unreachable; gateway or host down

// 被拒

R1#telnet 3.3.3.3

Trying 3.3.3.3 ... Open

// 可以

R3>exit

可以看到，R1 本地环回地址 192.168.1.1 无法远程访问 3.3.3.3，其他地址则可以。

R1#ping 8.8.8.8

Type escape sequence to abort.

Sending 5, 100-byte ICMP Echos to 8.8.8.8, timeout is 2 seconds:

!!!!!

// 成功

Success rate is 100 percent (5/5), round-trip min/avg/max = 36/40/44 ms

R1#ping 8.8.8.8 source loopback 2

Type escape sequence to abort.

Sending 5, 100-byte ICMP Echos to 8.8.8.8, timeout is 2 seconds:

Packet sent with a source address of 192.168.2.1

UUUUU

// 被拒

Success rate is 0 percent (0/5)

此时 192.168.2.0 网段无法 Ping 通 8.8.8.8，但是其他网段可以。

4. 管理命名 ACL。

①查看 ACL 状态，如下所示。

R2#show ip access-lists

// 查看 ACL 的状态

Extended IP access list DENYSERVICE

    10 deny tcp 192.168.1.0 0.0.0.255 host 3.3.3.3 eq telnet (3 matches)

//matches 表示多少次的匹配，10 是 ACL 列表的序号，从小到大的顺序

> 20 deny icmp 192.168.2.0 0.0.0.255 host 8.8.8.8 (15 matches)
>
> 30 permit ip any any (20 matches)

②删除 ACL 语句，如下所示。

> R2(config)#ip access-list extended DENYSERVICE
>
> // 进入 DENYSERVICE 这个 ACL 进程
>
> R2(config-ext-nacl)#no 10
>
> // 删除序号为 10 的语句

③查看 ACL 状态，如下所示。

> R2#show ip access-lists
>
> Extended IP access list DENYSERVICE
>
>   20 deny icmp 192.168.2.0 0.0.0.255 host 8.8.8.8 (15 matches)
>
>   30 permit ip any any (20 matches)

④插入 ACL 语句，如下所示。

> R2(config)#ip access-list extended DENYSERVICE
>
> // 进入 DENYSERVICE 这个 ACL 的进程
>
> R2(config-ext-nacl)#25 deny tcp 192.168.1.0 0.0.0.255 host 3.3.3.3 eq telnet
>
> // 新增一个序号为 25 的语句，在其位置在 20 和 30 之间

⑤查看 ACL 状态，如下所示。

> Extended IP access list DENYSERVICE
>
>   20 deny icmp 192.168.2.0 0.0.0.255 host 8.8.8.8 (15 matches)
>
>   25 deny tcp 192.168.1.0 0.0.0.255 host 3.3.3.3 eq telnet
>
>   30 permit ip any any (20 matches)

从上面实验可以看出，基于命名的标准和拓展 ACL，语法上与基于编号的 ACL 没有任何区别，只不过是将编号换成名字罢了，即便如此，命名 ACL 在管理方面更加体现人性化。此实验完成。

7.4 时间 ACL

实验目的：
1. 掌握时间 ACL 的基本编写。
2. 理解时间 ACL 的基本特性。

实验拓扑：

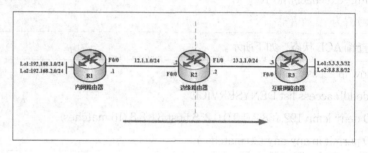

实验随手记：

_____

_____

_____

_____

_____

实验原理：

时间 ACL<Time-Based ACL> 是在原有 ACL 的基础上加入时间属性，可以实现更加弹性的访问控制。举个例子，我们可以在校园网出口路由器上执行 ACL，需求如下：本校区所有电脑不得上网！以往的 ACL 没有时间属性，则访问和拒绝是被固定化了。如果加入时间属性，则需求可以这么修改：本校区所有电脑在晚上 3 点到 5 点不得上网！这个需求就更加体现弹性化、人性化了。另一种使用环境，即企业环境，要求上班时间从早上 8 点半到下午 6 点不得上网，其他时间都可以，这也可以通过时间 ACL 来实现。所以时间 ACL 在实际工程环境下使用得非常多。

实验步骤：

1. 依据图中拓扑，配置各个路由器的 IP 地址，并部署静态路由保证全网连通。
R1 上配置如下所示。

R1(config)#ip route 23.1.1.0 255.255.255.0 12.1.1.2

R1(config)#ip route 3.3.3.3 255.255.255.255 12.1.1.2

R1(config)#ip route 8.8.8.8 255.255.255.255 12.1.1.2

R2 上配置如下所示。

```
R2(config)#ip route 192.168.1.0 255.255.255.0 12.1.1.1
R2(config)#ip route 192.168.2.0 255.255.255.0 12.1.1.1
R2(config)#ip route 3.3.3.3 255.255.255.255 23.1.1.3
R2(config)#ip route 8.8.8.8 255.255.255.255 23.1.1.3
```

R3 上配置如下所示。

```
R3(config)#ip route 12.1.1.0 255.255.255.0 23.1.1.2
R3(config)#ip route 192.168.1.0 255.255.255.0 23.1.1.2
R3(config)#ip route 192.168.2.0 255.255.255.0 23.1.1.2
```

测试连通性，如下所示。

```
R1#ping 3.3.3.3 source 192.168.1.1

Type escape sequence to abort.
Sending 5, 100-byte ICMP Echos to 3.3.3.3, timeout is 2 seconds:
Packet sent with a source address of 192.168.1.1
!!!!!
Success rate is 100 percent (5/5), round-trip min/avg/max = 40/40/40 ms
R1#ping 8.8.8.8 source 192.168.2.1

Type escape sequence to abort.
Sending 5, 100-byte ICMP Echos to 8.8.8.8, timeout is 2 seconds:
Packet sent with a source address of 192.168.2.1
!!!!!
Success rate is 100 percent (5/5), round-trip min/avg/max = 36/43/56 ms
```

可以看到，内网与外网通信没有问题。

2. 在 R2 上部署基于时间的 ACL，使得内网网段 192.168.1.0/24 在工作日上班时间 8:30 到 18:00 不能够访问外网。

①设备路由器本地时间，如下所示。

```
R2(config)#clock timezone BJ +8
// 定义时区
R2#clock set 10:00:00 30 JULY 2013
// 定义具体时间
```

②设备时间范围，如下所示。

```
R2(config)#time-range PL
// 为这个时间范围命名
R2(config-time-range)#periodic weekdays 8:30 to 18:00
// 时间范围设置有两种方式：
```

1. Periodic：周期时间；
2. Absolute：绝对时间。

周期时间能够"循环"，而绝对时间是"一次性"的

③设置 ACL 并调用时间范围，如下所示。

```
R2(config)#access-list 100 deny ip 192.168.1.0 0.0.0.255 any time-range PL
// 把这个时间方案加在最后。标准 ACL 不能加入时间范围，只有拓展 ACL 可以
R2(config)#access-list 100 permit ip any any
```

④接口下调用 ACL，如下所示。

```
R2(config)#int f0/0
R2(config-if)#ip access-group 100 in
```

3. 测试时间 ACL。

查看时间范围，如下所示。

```
R2#show time-range
time-range entry: PL (active)
// Active 表示此时间范围有效
    periodic weekdays 8:30 to 18:00
```

查看 ACL 状态，如下所示。

```
R2#show ip access-lists
Extended IP access list 100
    10 deny ip 192.168.1.0 0.0.0.255 any time-range PL (active)
    20 permit ip any any
```

让 R1 访问外网，如下所示。

```
R1#ping 8.8.8.8 source 192.168.1.1

Type escape sequence to abort.
Sending 5, 100-byte ICMP Echos to 8.8.8.8, timeout is 2 seconds:
Packet sent with a source address of 192.168.1.1
UUUUU
// 被拒
Success rate is 0 percent (0/5)
R1#ping 3.3.3.3 source 192.168.1.1

Type escape sequence to abort.
Sending 5, 100-byte ICMP Echos to 3.3.3.3, timeout is 2 seconds:
Packet sent with a source address of 192.168.1.1
UUUUU
```

```
Success rate is 0 percent (0/5)
```

可以看到，在特定时间范围内，此时 192.168.1.0 网段无法访问外网。可以尝试将本地路由器的时间修改。

修改本地时间，如下所示。

```
R2#clock set 22:00:00 30 JULY 2013
```

查看时间范围，如下所示。

```
R2#show time-range
time-range entry: PL (inactive)
//Inactive 表示此时间范围失效
periodic weekdays 8:30 to 18:00
    used in: IP ACL entry
```

查看 ACL 状态，如下所示。

```
R2#show ip access-lists
Extended IP access list 100
    10 deny ip 192.168.1.0 0.0.0.255 any time-range PL (inactive) (30 matches)
    20 permit ip any any
```

让 R1 访问外网，如下所示。

```
R1#ping 8.8.8.8 source 192.168.1.1

Type escape sequence to abort.
Sending 5, 100-byte ICMP Echos to 8.8.8.8, timeout is 2 seconds:
Packet sent with a source address of 192.168.1.1
!!!!!
Success rate is 100 percent (5/5), round-trip min/avg/max = 48/60/64 ms
R1#ping 3.3.3.3 source 192.168.1.1

Type escape sequence to abort.
Sending 5, 100-byte ICMP Echos to 3.3.3.3, timeout is 2 seconds:
Packet sent with a source address of 192.168.1.1
!!!!!
Success rate is 100 percent (5/5), round-trip min/avg/max = 52/63/68 ms
```

可以看到，当不在特定的时间范围内，192.168.1.0/24 能够访问外网。

从以上实验可以看出，时间 ACL 相比传统的 ACL 来说，加入了时间属性，使得流量管理更加弹性化。此实验完成。

7.5 动态 NAT

实验目的：
1. 掌握动态 NAT 的基本编写。
2. 理解动态 NAT 的基本特性。

实验拓扑：

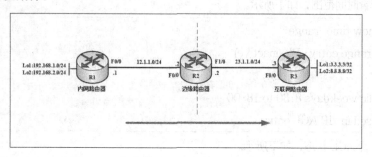

实验随手记：

_____
_____
_____
_____
_____
_____
_____

实验原理：

1. NAT 概述

NAT（Network Address Translation，网络地址转换）技术可以用于实现地址翻转，节省地址使用并有效隐藏内网实现安全。在目前地址紧缺的环境下，NAT 是局域网连接到互联网时必须要部署的"标配"。

2. NAT 分类

根据地址翻转的特征，可分为三种 NAT 技术，如下表所示。

NAT 类型	功能	特征
动态 NAT	Dynamic NAT（DAT），用于实现多对多的动态地址映射	需要购买大量的地址，用于实现内网访问互联网，比较消耗地址资源
端口 NAT	Port NAT（PAT），用于实现多对少的地址复用，是目前使用最为广泛的 NAT 地址	只需要购买少量地址，用于实现内网访问互联网，可以将若干地址复用给内网的所有主机
静态 NAT	Static NAT（SAT），用于实现一对一的地址映射	一般用于将内网服务器映射到互联网，使得互联网用户可以访问本地服务器，根据服务器需求购买地址

7.5 动态 NAT

实验步骤:

1. 依据图中拓扑,配置各个路由器的 IP 地址,并部署静态和默认路由。

R1 上配置如下所示。

R1(config)# ip route 0.0.0.0 0.0.0.0 12.1.1.2
// 默认路由,指向 R2(边缘路由器)

R2 上配置如下所示。

R2(config)#ip route 192.168.1.0 255.255.255.0 12.1.1.1
// 静态路由,指回 R1
R2(config)#ip route 192.168.2.0 255.255.255.0 12.1.1.1
R2(config)#ip route 0.0.0.0 0.0.0.0 100.1.23.3
// 默认路由器,指向 R3

测试连通性,如下所示。

R2#ping 8.8.8.8

Type escape sequence to abort.
Sending 5, 100-byte ICMP Echos to 8.8.8.8, timeout is 2 seconds:
!!!!!
Success rate is 100 percent (5/5), round-trip min/avg/max = 28/33/44 ms
R1#ping 8.8.8.8 source 192.168.1.1

Type escape sequence to abort.
Sending 5, 100-byte ICMP Echos to 8.8.8.8, timeout is 2 seconds:
Packet sent with a source address of 192.168.1.1
.....
// 失败了。可以出去,不能回来,想想为什么?
Success rate is 0 percent (0/5)

可以看到,边缘路由器可以访问互联网,但是内网无法访问互联网。因为互联网路由器没有到内网的私有路由,一般需要部署 NAT 技术,将内网地址转换成公网地址,才能实现通信。

2. 在 R2 上部署动态 NAT,使得内网所有设备可以访问外网。

① 定义内网流量和公网地址池,如下所示。

R2(config)#access-list 1 permit 192.168.1.0 0.0.0.255
// 抓住感兴趣的流量
R2(config)#access-list 2 permit 192.168.2.0 0.0.0.255
R2(config)#ip nat pool DNAT 100.1.23.100 100.1.23.200 netmask 255.255.255.0
//DNAT 是该 NAT 的名字,地址范围是从 100.1.23.100 到 100.1.23.200

② 定义内外接口,如下所示。

```
R2(config)#int f0/0
R2(config-if)#ip nat inside
// 把 f0/0 作为 NAT 的内网端口
R2(config-if)#exit
R2(config)#int f1/0
R2(config-if)#ip nat outside
// 把 f1/0 作为 NAT 的外部端口
R2(config-if)#exit
```

③执行动态 NAT，如下所示。

```
R2(config)#ip nat inside source list 1 pool DNAT
// 执行！
```

3. 测试动态 NAT。

在 R2 上调试 NAT 进程，如下所示。

```
R2#debug ip nat
// 查看所有有关 NAT 的数据包
```

在 R1 上访问互联网，如下所示。

```
R1#ping 8.8.8.8 source 192.168.1.1

Type escape sequence to abort.
Sending 5, 100-byte ICMP Echos to 8.8.8.8, timeout is 2 seconds:
Packet sent with a source address of 192.168.1.1
.!!!!
Success rate is 80 percent (4/5), round-trip min/avg/max = 60/77/124 ms
R1#ping 8.8.8.8 source 192.168.2.1

Type escape sequence to abort.
Sending 5, 100-byte ICMP Echos to 8.8.8.8, timeout is 2 seconds:
Packet sent with a source address of 192.168.2.1
.!!!!
Success rate is 80 percent (4/5), round-trip min/avg/max = 60/69/76 ms
```

从上面可以看到内网已经能够访问外网。

在 R2 上查看 NAT 调试信息，如下所示。

```
R2#
*Mar  1 00:27:01.063: NAT*: s=192.168.1.1->100.1.23.100, d=8.8.8.8 [6]
*Mar  1 00:27:01.127: NAT*: s=8.8.8.8, d=100.1.23.100->192.168.1.1 [6]
```

```
*Mar  1 00:27:01.155: NAT*: s=192.168.1.1->100.1.23.100, d=8.8.8.8 [7]
*Mar  1 00:27:01.187: NAT*: s=8.8.8.8, d=100.1.23.100->192.168.1.1 [7]
*Mar  1 00:27:01.219: NAT*: s=192.168.1.1->100.1.23.100, d=8.8.8.8 [8]
*Mar  1 00:27:01.247: NAT*: s=8.8.8.8, d=100.1.23.100->192.168.1.1 [8]
*Mar  1 00:27:01.279: NAT*: s=192.168.1.1->100.1.23.100, d=8.8.8.8 [9]
*Mar  1 00:27:01.307: NAT*: s=8.8.8.8, d=100.1.23.100->192.168.1.1 [9]

*Mar  1 00:27:44.667: NAT*: s=192.168.2.1->100.1.23.101, d=8.8.8.8 [16]
*Mar  1 00:27:44.719: NAT*: s=8.8.8.8, d=100.1.23.101->192.168.2.1 [16]
*Mar  1 00:27:44.751: NAT*: s=192.168.2.1->100.1.23.101, d=8.8.8.8 [17]
*Mar  1 00:27:44.783: NAT*: s=8.8.8.8, d=100.1.23.101->192.168.2.1 [17]
*Mar  1 00:27:44.815: NAT*: s=192.168.2.1->100.1.23.101, d=8.8.8.8 [18]
*Mar  1 00:27:44.847: NAT*: s=8.8.8.8, d=100.1.23.101->192.168.2.1 [18]
*Mar  1 00:27:44.879: NAT*: s=192.168.2.1->100.1.23.101, d=8.8.8.8 [19]
*Mar  1 00:27:44.907: NAT*: s=8.8.8.8, d=100.1.23.101->192.168.2.1 [19]
// 可以清楚地看到地址转换的过程
```

从调试过程可以看到，不同的内网地址被翻转到不同的外部地址。

查看 R2 上 NAT 转换表，如下所示。

```
R2#show ip nat translations
// 查看 NAT 的转换映射
Pro Inside global      Inside local     Outside local    Outside global
icmp 100.1.23.100:4    192.168.1.1:4    8.8.8.8:4        8.8.8.8:4
---- 100.1.23.100      192.168.1.1      ---              ---
icmp 100.1.23.101:5    192.168.2.1:5    8.8.8.8:5        8.8.8.8:5
---- 100.1.23.101      192.168.2.1      ---              ---
----
```

从转换表可以看出，内网地址被一一映射到公网地址上。

查看 R2 上 NAT 转换状态，如下所示。

```
R2#show ip nat statistics
Total active translations: 2 (0 static, 2 dynamic; 0 extended)
Outside interfaces:
 FastEthernet1/0
Inside interfaces:
 FastEthernet0/0, FastEthernet3/0
Hits: 34  Misses: 4
CEF Translated packets: 38, CEF Punted packets: 0
Expired translations: 4
```

```
Dynamic mappings:
-- Inside Source
[Id: 2] access-list 1 pool DNAT refcount 2
 pool DNAT: netmask 255.255.255.0
     start 100.1.23.100 end 100.1.23.200
     type generic, total addresses 101, allocated 2 (1%), misses 0
Queued Packets: 0
```

从上面可以看到公网地址池分配的状态。通过本实验可以得到，动态 NAT 可以实现内网到外网的地址转换，并且将地址一对一映射出去。但是在 IPv4 地址数量不足的情况下，此解决方案并不能节省地址使用，因为每个私有地址需要对应一个公有地址。而后续的端口复用技术便可以使多个私有地址映射到一个公有地址，能够满足更加实际的工程需求。此实验完成。

7.6 端口 NAT

实验目的：
1. 掌握端口 NAT 的基本编写。
2. 理解端口 NAT 的基本特性。

实验拓扑：

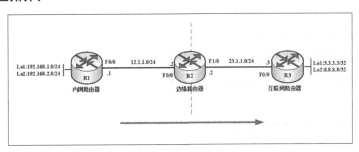

实验随手记：

实验步骤：
1. 依据图中拓扑，配置各个路由器的 IP 地址，并部署静态和默认路由。
   R1 上配置如下所示。

```
R1(config)# ip route 0.0.0.0 0.0.0.0 12.1.1.2
// 默认路由，指向 R2（边缘路由器）
```

   R2 上配置如下所示。

```
R2(config)#ip route 192.168.1.0 255.255.255.0 12.1.1.1
// 静态路由，指回 R1
R2(config)#ip route 192.168.2.0 255.255.255.0 12.1.1.1
R2(config)#ip route 0.0.0.0 0.0.0.0 100.1.23.3
// 默认路由器，指向 R3
```

   测试连通性，如下所示。

```
R2#ping 8.8.8.8
```

```
Type escape sequence to abort.
Sending 5, 100-byte ICMP Echos to 8.8.8.8, timeout is 2 seconds:
!!!!!
Success rate is 100 percent (5/5), round-trip min/avg/max = 28/33/44 ms

R1#ping 8.8.8.8 source 192.168.1.1

Type escape sequence to abort.
Sending 5, 100-byte ICMP Echos to 8.8.8.8, timeout is 2 seconds:
Packet sent with a source address of 192.168.1.1
……
Success rate is 0 percent (0/5)
```

可以看到,边缘路由器可以访问互联网,但是内网无法访问互联网。因为互联网路由器没有到内网的私有路由,一般需要部署 NAT 技术,将内网地址转换成公网地址,才能实现通信。

2. 在 R2 上部署端口 NAT,使得内网所有设备可以访问外网。

① 定义内网流量和公网地址池,如下所示。

```
R2(config)#access-list 1 permit 192.168.1.0 0.0.0.255
// 抓去感兴趣的流量
R2(config)#access-list 2 permit 192.168.2.0 0.0.0.255
R2(config)#ip nat pool PAT 100.1.23.100 100.1.23.101 netmask 255.255.255.0
//PAT 是该 NAT 的名字地址范围是从 100.1.23.100 到 100.1.23.101
```

② 定义内外接口,如下所示。

```
R2(config)#int f0/0
R2(config-if)#ip nat inside
R2(config-if)#exit
R2(config)#int f1/0
R2(config-if)#ip nat outside
R2(config-if)#exit
```

③ 执行端口 NAT

```
R2(config)#ip nat inside source list 1 pool PAT overload
//overload 表示重载和复用,端口复用
```

3. 测试端口 NAT。

在 R2 上调试 NAT 进程,如下所示。

```
R2#debug ip nat
```

在 R1 上访问互联网,如下所示。

```
R1#ping 8.8.8.8 source 192.168.1.1

Type escape sequence to abort.
Sending 5, 100-byte ICMP Echos to 8.8.8.8, timeout is 2 seconds:
Packet sent with a source address of 192.168.1.1
.!!!!
Success rate is 80 percent (4/5), round-trip min/avg/max = 60/77/124 ms
R1#ping 8.8.8.8 source 192.168.2.1

Type escape sequence to abort.
Sending 5, 100-byte ICMP Echos to 8.8.8.8, timeout is 2 seconds:
Packet sent with a source address of 192.168.2.1
.!!!!
Success rate is 80 percent (4/5), round-trip min/avg/max = 60/69/76 ms
```

从上面可以看到内网已经能够访问外网。

在 R2 上查看 NAT 调试信息，如下所示。

```
R2#
*Mar  1 01:06:13.027: NAT*: s=192.168.1.1->100.1.23.100, d=8.8.8.8 [30]
*Mar  1 01:06:13.083: NAT*: s=8.8.8.8, d=100.1.23.100->192.168.1.1 [30]
*Mar  1 01:06:13.119: NAT*: s=192.168.1.1->100.1.23.100, d=8.8.8.8 [31]
*Mar  1 01:06:13.163: NAT*: s=8.8.8.8, d=100.1.23.100->192.168.1.1 [31]
*Mar  1 01:06:13.195: NAT*: s=192.168.1.1->100.1.23.100, d=8.8.8.8 [32]
*Mar  1 01:06:13.223: NAT*: s=8.8.8.8, d=100.1.23.100->192.168.1.1 [32]
*Mar  1 01:06:13.255: NAT*: s=192.168.1.1->100.1.23.100, d=8.8.8.8 [33]
*Mar  1 01:06:13.287: NAT*: s=8.8.8.8, d=100.1.23.100->192.168.1.1 [33]
*Mar  1 01:06:13.319: NAT*: s=192.168.1.1->100.1.23.100, d=8.8.8.8 [34]
*Mar  1 01:06:13.351: NAT*: s=8.8.8.8, d=100.1.23.100->192.168.1.1 [34]
*Mar  1 01:06:15.131: NAT*: s=192.168.2.1->100.1.23.100, d=8.8.8.8 [35]
*Mar  1 01:06:15.163: NAT*: s=8.8.8.8, d=100.1.23.100->192.168.2.1 [35]
*Mar  1 01:06:15.207: NAT*: s=192.168.2.1->100.1.23.100, d=8.8.8.8 [36]
*Mar  1 01:06:15.255: NAT*: s=8.8.8.8, d=100.1.23.100->192.168.2.1 [36]
*Mar  1 01:06:15.303: NAT*: s=192.168.2.1->100.1.23.100, d=8.8.8.8 [37]
*Mar  1 01:06:15.335: NAT*: s=8.8.8.8, d=100.1.23.100->192.168.2.1 [37]
*Mar  1 01:06:15.367: NAT*: s=192.168.2.1->100.1.23.100, d=8.8.8.8 [38]
*Mar  1 01:06:15.399: NAT*: s=8.8.8.8, d=100.1.23.100->192.168.2.1 [38]
*Mar  1 01:06:15.427: NAT*: s=192.168.2.1->100.1.23.100, d=8.8.8.8 [39]
*Mar  1 01:06:15.459: NAT*: s=8.8.8.8, d=100.1.23.100->192.168.2.1 [39]
```

查看 R2 上 NAT 转换表，如下所示。

```
R2#show ip nat translations
Pro Inside global        Inside local       Outside local      Outside global
icmp 100.1.23.100:6      192.168.1.1:6      8.8.8.8:6          8.8.8.8:6
icmp 100.1.23.100:7      192.168.2.1:7      8.8.8.8:7          8.8.8.8:7
```

从上面可以看到，当采用端口复用技术时，多个内网地址被映射到同一个公网地址上，实现多对一映射。当有多个公网地址时，需要采用地址池方式进行地址复用。若公网地址只有一个例如出接口地址时，则只需调用出接口进行地址复用。

4. 通过接口部署端口 NAT。

①定义内网流量，如下所示。

```
R2(config)#access-list 1 permit 192.168.1.0 0.0.0.255
R2(config)#access-list 2 permit 192.168.2.0 0.0.0.255
```

②定义内外接口，如下所示。

```
R2(config)#int f0/0
R2(config-if)#ip nat inside
R2(config-if)#exit
R2(config)#int f1/0
R2(config-if)#ip nat outside
R2(config-if)#exit
```

③执行端口 NAT，如下所示。

```
R2(config)#ip nat inside source list 1 interface f1/0 overload
// 注意看，这个不是 POOL 地址池了，而直接是一个端口，即直接转成端口上的 IP 地址
```

在 R1 上访问互联网，如下所示。

```
R1#ping 8.8.8.8 source 192.168.1.1

Type escape sequence to abort.
Sending 5, 100-byte ICMP Echos to 8.8.8.8, timeout is 2 seconds:
Packet sent with a source address of 192.168.1.1
.!!!!
Success rate is 80 percent (4/5), round-trip min/avg/max = 60/77/124 ms
R1#ping 8.8.8.8 source 192.168.2.1

Type escape sequence to abort.
Sending 5, 100-byte ICMP Echos to 8.8.8.8, timeout is 2 seconds:
Packet sent with a source address of 192.168.2.1
.!!!!
```

Success rate is 80 percent (4/5), round-trip min/avg/max = 60/69/76 ms

在 R2 上查看 NAT 转换表，如下所示。

```
R2#show ip nat translations
Pro Inside global      Inside local      Outside local    Outside global
icmp 100.1.23.2:8      192.168.1.1:8     8.8.8.8:8        8.8.8.8:8
icmp 100.1.23.2:9      192.168.2.1:9     8.8.8.8:9        8.8.8.8:9
```

此时，多个内网地址被映射到同一个公网出口地址。通过以上实验可以得到，不管是采用地址池还是接口进行地址复用，都可以实现少对多甚至一对多的映射。这样的话，在公网地址不足的情况下便可以很好地节省地址资源，相比动态 NAT 来说，更适合一般企业的应用。此实验完成。

7.7 静态 NAT

实验目的：
1. 掌握静态 NAT 的基本编写。
2. 理解静态 NAT 的基本特性。

实验拓扑：

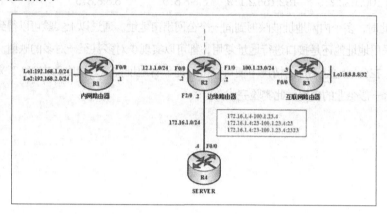

实验随手记：

_____
_____
_____
_____
_____
_____
_____

实验步骤：

1. 依据图中拓扑，配置各个路由器的 IP 地址，并部署静态和默认路由。

R1 上配置如下所示。

R1(config)# ip route 0.0.0.0 0.0.0.0 12.1.1.2
// 默认路由，指向 R2（边缘路由器）

R2 上配置如下所示。

R2(config)#ip route 192.168.1.0 255.255.255.0 12.1.1.1
// 静态路由，指回 R1
R2(config)#ip route 192.168.2.0 255.255.255.0 12.1.1.1
R2(config)#ip route 0.0.0.0 0.0.0.0 100.1.23.3
// 默认路由器，指向 R3

7.7 静态 NAT

R4 上配置如下所示。

```
R4(config)#ip route 0.0.0.0 0.0.0.0 172.16.1.2
// 默认路由，指向 R2
R4(config)#line vty 0 15
R4(config-line)#no login
// 无需密码验证即可登录，方便实验
R4(config-line)#privilege level 15
// 最高权限，一进去就是特权模式
R4(config-line)#exit
```

让互联网设备访问内网服务器，如下所示。

```
R3#ping 172.16.1.4

Type escape sequence to abort.
Sending 5, 100-byte ICMP Echos to 172.16.1.4, timeout is 2 seconds:
.....
// 不通
Success rate is 0 percent (0/5)
```

毫无疑问，由于一般的服务器搭建在企业内网或者 IDC 数据中心机房，没有直接放置在互联网上，互联网主机之间没法访问到内网的服务器，而静态 NAT 便是用来解决这个问题。通过部署静态 NAT，可以将服务器映射到公网，使公网主机可以访问。

2. 在 R2 上部署静态 NAT。

①定义内外接口，如下所示。

```
R2(config)#int f2/0
R2(config-if)#ip nat inside
R2(config-if)#exit
R2(config)#int f1/0
R2(config-if)#ip nat outside
R2(config-if)#exit
```

②执行静态 NAT，如下所示。

```
R2(config)#ip nat inside source static 172.16.1.4 100.1.23.4
//Static 是静态 NAT 的关键字，后面直接映射两个地址
```

3. 测试端口 NAT。

在 R2 上查看 NAT 转换表，如下所示。

```
R2#show ip nat translations
Pro Inside global    Inside local    Outside local    Outside global
--- 100.1.23.4       172.16.1.4      ---
---
```

此时让 R3 访问 R4，如下所示。

```
R3#ping 100.1.23.4

Type escape sequence to abort.
Sending 5, 100-byte ICMP Echos to 100.1.23.4, timeout is 2 seconds:
!!!!!
Success rate is 100 percent (5/5), round-trip min/avg/max = 56/72/112 ms
R3#telnet 100.1.23.4
Trying 100.1.23.4 ... Open
// 成功
R4#
```

可以看到，通过部署静态 NAT，互联网主机可以通过映射后的公网地址访问内网服务器。上面这种静态映射是直接将内部 IP 映射到外部 IP，这样相当于将内网服务器的所有端口都开放到互联网上，非常不安全，一般需要映射到具体的端口上。

4. 部署静态 NAT，将 R4 的 23 端口映射到公网上，如下所示。

```
R2(config)#ip nat inside source static tcp 172.16.1.4 23 100.1.23.4 23
// 还可以加上对应的端口号
```

此映射方法只将本地服务 23 端口放开，保证安全性。当然，还有更加安全隐藏的映射方式，例如本地服务端口和映射端口可以不一致。

5. 部署静态 NAT，将 R4 的 23 端口通过 2323 端口映射到公网上，如下所示。

```
R2(config)#ip nat inside source static tcp 172.16.1.4 23 100.1.23.4 2323
// 把端口 23 对外网映射成 2323，去制造迷惑，产生安全
```

此时让 R3 通过 2323 端口远程访问 R4，如下所示。

```
R3#telnet 100.1.23.4 2323
Trying 100.1.23.4, 2323 ... Open

R4#
```

若只有一个公网地址，却有多个内部服务器时，静态 NAT 同样可以做到地址复用。

6. 部署静态 NAT，将公网地址映射给多个服务器，如下所示。

```
R2(config)#ip nat inside source static tcp 172.16.1.4 23 100.1.23.4 23
R2(config)#ip nat inside source static tcp 172.16.1.5 80 100.1.23.4 80
R2(config)#ip nat inside source static tcp 172.16.1.6 443 100.1.23.4 443
// 一址多用
```

查看 R2 上转换表，如下所示。

```
R2#show ip nat translations
Pro Inside global     Inside local      Outside local    Outside global
tcp 100.1.23.4:23     172.16.1.4:23     ---              ---
tcp 100.1.23.4:80     172.16.1.5:80     ---              ---
tcp 100.1.23.4:443    172.16.1.6:443    ---              ---
```

可以看到，此时通过不同端口，多个服务器映射到同一个公网地址。此实验完成。

第 8 章 高级安全

本章主要学习高级安全技术，包括防火墙和 VPN 技术。防火墙是园区网络边缘环境下的安全设备，为整个网络提供防护攻击，防止外部黑客攻击。本章通过图形化工具 ASDM 软件对思科防火墙产品进行调试。除了防火墙技术，本章还将学习 VPN 技术，VPN 技术是不同于局域网站点用来实现安全通信的逻辑隧道技术，可以保证数据经过互联网的时候不被窃取和修改，是一种数据安全技术。以下为本章导航图。

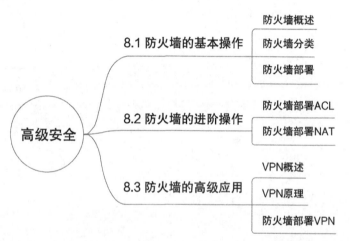

8.1 防火墙的基本操作

实验目的：
1. 掌握 SDM 图形化软件的部署。
2. 掌握 SDM 图形化 IOS 的配置。
3. 掌握 ASDM 图形化 PIX 的配置。

实验拓扑：

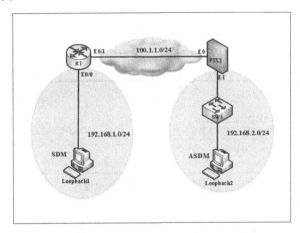

实验随手记：

_____
_____
_____
_____
_____
_____

实验原理：

1. 防火墙概述

防火墙是网络中用于实现安全防护的最基础的设备。防火墙主要有三种分类，包括包过滤防火墙、应用代理防火墙、状态化过滤防火墙。以下介绍这三种不同防火墙的工作原理。

①包过滤防火墙：类似访问控制列表，只能对数据包的 3、4 层进行过滤，没法进行数据的深度过滤，功能比较弱。

②应用代理防火墙：类似代理服务器，内部主机与外部通信时，将数据发送给应用代理防火墙,防火墙根据本地的安全规则进行过滤,需要占用的资源比较大,速度较慢。

③状态化过滤防火墙：目前主流防火墙的工作原理，内部主机访问外部时，防火

墙根据自适应算法，对数据包进行状态会话记录，生成会话记录，当数据从外部返回时，根据状态表匹配会话并返回给内部主机。若没有匹配，则不通过状态化过滤防火墙如图 8-1 所示。

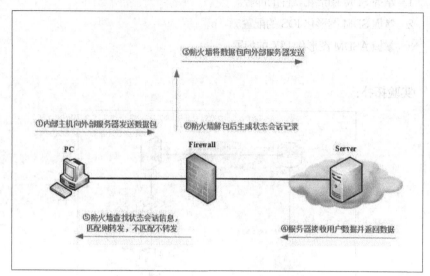

图 8-1　状态化过滤防火墙

2．防火墙部署

防火墙一般部署在网络的边界，实现区域间安全防护。图 8-2 是常规的防火墙网络拓扑设计，INSIDE 区域一般连接内网；DMZ 区域称为非军事化区域，连接内部服务器；INTERNET 区域连接外部网络。内部区域安全级别最高，互联网区域最低，级别高的区域可以主动访问级别低的区域，反之不行。

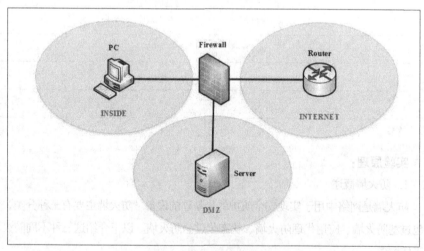

图 8-2　防火墙部署拓扑

实验步骤：

1．在电脑上建立两个环回接口，用来桥接到 GNS3 的网络，具体方法参考（本书第 1 章 GNS 安装与使用）。我们电脑的 Loopback1 桥接到 R1，Loopback2 桥接到 PIX1，Loopback2 和 PIX 之间不能相连，需要通过普通的以太网交换机来连接。这里我们选用 GNS3 自带的以太网交换机，R1 选用 NM-4E 模块。

2. 配置 R1 相连的 Loopback1 的地址，如图 8-3 所示。

图 8-3　电脑环回口 IP 配置

R1 上配置，如下所示。

```
R1(config)#interface e0/0
R1(config-if)#no shutdown
R1(config-if)#ip add 192.168.1.254 255.255.255.0
R1(config-if)#interface e0/1
R1(config-if)#no shutdown
R1(config-if)#ip add 100.1.1.1 255.255.255.0
```

测试连通性，如图 8-4 所示。

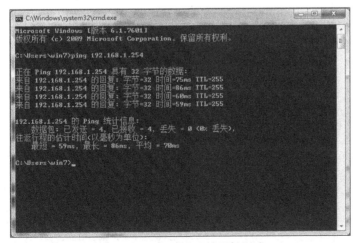

图 8-4　环回口与 R1 连通性测试

3. 安装 SDM 相关的软件（JRE，SDM）。

（1）先安装 JRE（Java 虚拟机），如图 8-5 和图 8-6 所示。

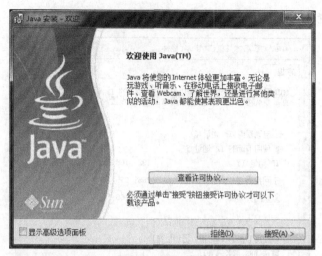

图 8-5　Java 安装

图 8-6　Java 安装 2

（2）再安装 SDM 软件，如图 8-7、图 8-8 和图 8-9 所示。

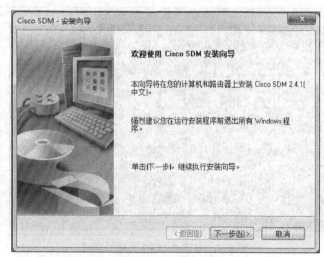

图 8-7　SDM 安装

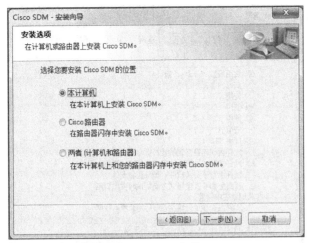

图 8-8 SDM 安装 2

图 8-9 SDM 安装 3

4. 配置 R1 与 SDM 连接前的准备工作，如下所示。

R1(config)#ip http server

// 开启 http 服务

R1(config)#ip http secure-server

// 开启 https 服务

% Generating 1024 bit RSA keys, keys will be non-exportable...[OK]

R1(config)#ip http authentication local

// 调用本地数据库认证

R1(config)#username test privilege 15 password test

// 创建用户数据库，并给其 15 级权限

R1(config)#line vty 0 4

R1(config-line)#login local

5. 设置默认浏览器为 IE，打开 IE 浏览器，"工具"→"Internet 选项"，打开 "高级"选项卡，勾选"允许活动内容在我的计算机上运行 *"，单击"确定"按钮，如图 8-10 所示。

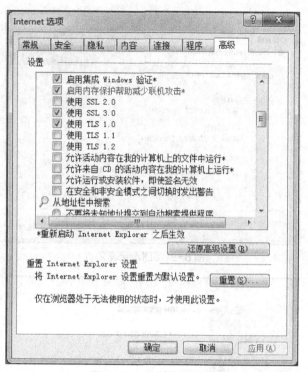

图 8-10 IE 浏览器设置

6. 单击桌面上的"Cisco SDM",输入路由器的 IP 地址,单击"启动"按钮,如图 8-11 所示。

弹出 IE,选择允许运行,如图 8-12 所示。

图 8-11 SDM 操作

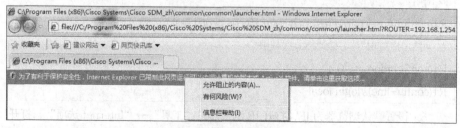

图 8-12 SDM 操作 2

输入我们配置的用户名和密码,之后会弹出警告,勾选"始终信任此发行者的内容",单击"运行"按钮,如图 8-13 所示。

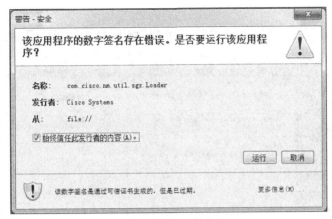

图 8-13 SDM 操作 3

之后还需再次输入用户名和密码,就进入了 SDM 的主界面,如图 8-14 和图 8-15 所示。

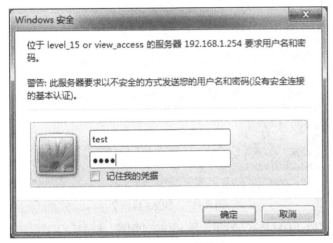

图 8-14 SDM 操作 4

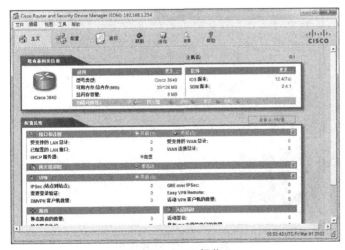

图 8-15 SDM 操作 5

7. 用 SDM 配置 NAT,如图 8-16 所示。依次单击"配置"→"NAT"→"基

本 NAT"→"启动选定的任务",此时会弹出 NAT 的配置向导,如图 8-17 所示。

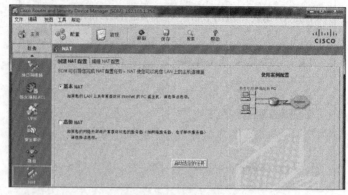

图 8-16　SDM 操作 6

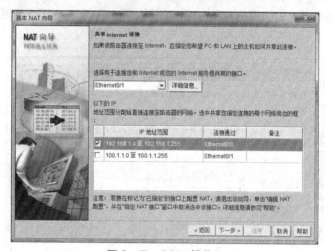

图 8-17　SDM 操作 7

单击结束后,SDM 会往 IOS 里面导入刚刚的配置,此时 NAT 就做完了,如图 8-18 所示。

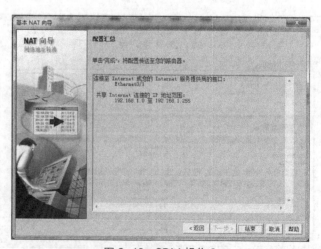

图 8-18　SDM 操作 8

8. 部署 ASDM,首先配置 Loopback2 的 IP,如图 8-19 所示。

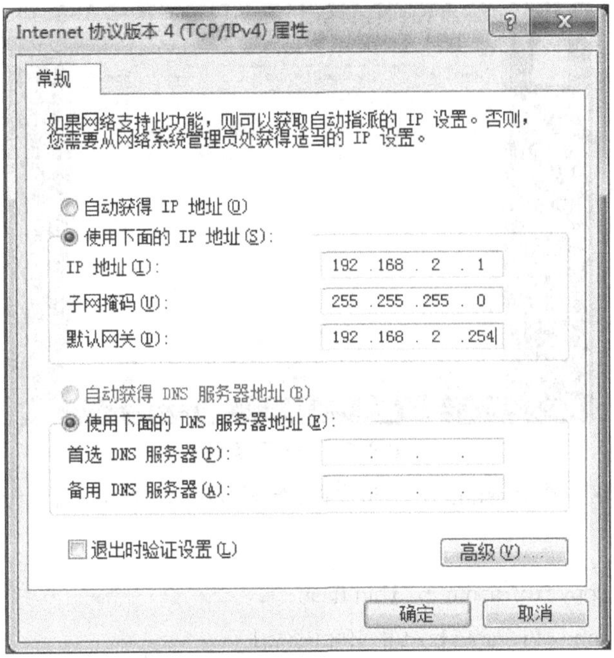

图 8-19　电脑环回口的 IP 配置

9. 配置 PIX 的 IP，如下所示。

```
pixfirewall(config)# fixup protocol icmp
// 关闭防火墙对 ICMP 的检查，测试时需开启
pixfirewall(config)# interface e1
pixfirewall(config-if)#no shutdown
pixfirewall(config-if)# ip add 192.168.2.254 255.255.255.0
pixfirewall(config-if)# nameif inside// 指定为内部区域
INFO: Security level for "inside" set to 100 by default.
```

测试连通性，如图 8-20 所示。

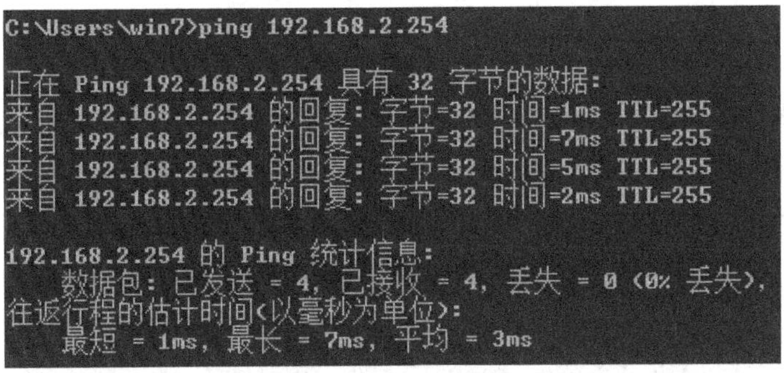

图 8-20　连通性测试

10. 打开 tftp 软件，将事先准备好的 asdm-613.bin 和 tftp 放在同一目录下，如图 8-21 所示。

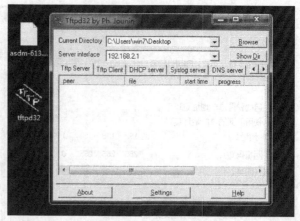

图 8-21　TFTP 安装

11. 在 PIX 中拷贝 asdm-614.bin 文件，需要 5~10 分钟，可以在 tftp 中查看进度，如下所示。

pixfirewall# copy tftp:asdm-613.bin flash:

Address or name of remote host []? 192.168.2.1

Source filename [asdm-613.bin]?　　　// 回车

Destination filename [asdm-613.bin]?　　// 回车

12. PIX 开启 ASDM，如下所示。

pixfirewall(config)# http server enable

// 启用 http 服务

pixfirewall(config)# http 192.168.2.0 255.255.255.0 inside

// 设置允许访问 asdm 的主机，其中 inside 为接口名

pixfirewall(config)# username test password test privilege 15

// 设置用户名和密码，注意一定要有 15 级权限

pixfirewall(config)# asdm image flash:/asdm-613.bin

// 指定 asdm 镜像的位置

13. 在 IE 浏览器中输入 https://192.168.2.254，其中 192.168.2.254 为 pix 的 IP，之后弹出证书错误，选择"继续浏览此网站"。

14. 单击 InstallASDM Laucher and Run ASDM，输入用户名和密码，之后会自动下载 ASDM Launcher 并安装，如图 8-22 所示。

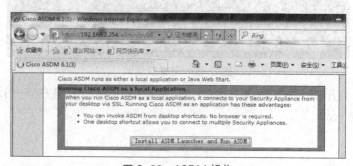

图 8-22　ASDM 操作

15. 打开 ADSM Launcher，输入 PIX 的地址和用户名、密码，单击"OK"按钮，如图 8-23 所示，再在弹出的对话框中单击"运行"按钮。

图 8-23　ASDM 操作

16. ASDM 的主界面和 SDM 的主界面相似，单击"configuration→interface"，双击"Ethernet 0"，配置该接口 IP，单击"OK"按钮，如图 8-24 所示。

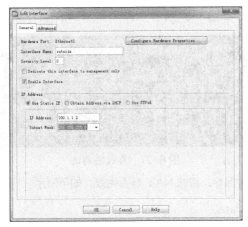

图 8-24　ASDM 操作

17. 为 PIX 配置 NAT，依次勾选"Configuration"→"Firewall"→"NAT Rules"，单击"Add"按钮，如图 8-25 所示。

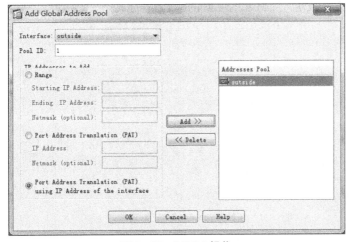

图 8-25　ASDM 操作

Manage→Add，Interface→Outside，勾选"Port Address Translation"，Add→OK，如图 8-26 所示。

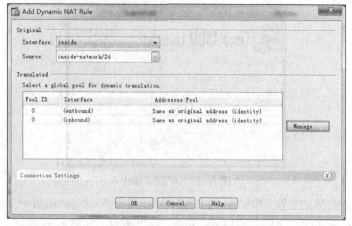

图 8-26　ASDM 操作

18. 测试连通性，如图 8-27 所示。

图 8-27　连通性测试

19. Apply 应用并保存，测试 NAT 是否生效，如下所示。

pixfirewall# show xlate

// 查看 NAT 转换表

1 in use, 1 most used

PAT Global 100.1.1.2(10098) Local 192.168.2.1 ICMP id 1

NAT 转换成功，SDM 和 ASDM 的基本操作完成！

8.2 防火墙的进阶操作

实验目的:
1. 通过 ASDM 部署 ACL。
2. 通过 ASDM 部署静态 NAT。
3. 通过 ASDM 部署端口 NAT。

实验拓扑:

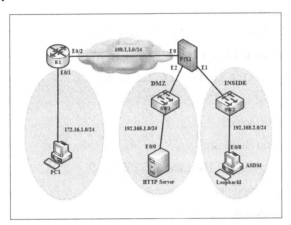

实验随手记:

_____
_____
_____
_____
_____
_____

实验说明:

1. PIX 为某企业的边缘防火墙,R1 为该企业分支部门的边缘路由器,现在要实现 http server、PC1 和 PC2 都能够上互联网,实现分支 PC1 访问总部的 http 服务,且映射 80 端口到端口 8008 来隐藏服务器的真实端口(提示:通过静态 NAT 实现)。

2. PC1、http server、R1 用路由器 3640 来模拟,且都加 NM-4E 模块,中间的云来代表互联网,网段为 100.1.1.0/24,左边的交换机用 GNS3 自带的以太网交换机,把真机的 Loopback1 桥接到 GNS3 中模拟分支 PC。

实验步骤:

1. 公司总部配置基础 IP 地址。
PIX 配置如下所示。

```
PIX1(config)# interface e0
PIX1(config-if)# no shutdown
PIX1(config-if)# ip add 100.1.1.1 255.255.255.0
PIX1(config-if)# nameif outside
INFO: Security level for "outside" set to 0 by default.
// 设置此接口为 outside，它的默认安全级别为 0
PIX1(config)# interface e1
PIX1(config-if)# no shutdown
PIX1(config-if)# ip add 192.168.2.254 255.255.255.0
PIX1(config-if)# nameif inside
INFO: Security level for "inside" set to 100 by default.
// 设置此接口为 inside，它的默认安全级别为 100
PIX1(config)# interface e2
PIX1(config-if)# no shutdown
PIX1(config-if)# ip add 192.168.1.254 255.255.255.0
PIX1(config-if)# nameif dmz
PIX1(config-if)# security-level 50
// 手动指定 dmz 的安全级别为 50，防止 outside 等安全级别低的访问
```

http server 配置如下所示。

```
http_server(config)#interface e0/0
http_server(config-if)#no shutdown
http_server(config-if)#ip add 192.168.1.1 255.255.255.0
http_server(config)#no ip routing
http_server(config)#ip default-gateway 192.168.1.254
```

真机配置如图 8-28 所示。

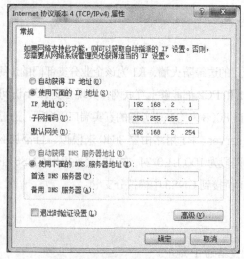

图 8-28　电脑环回口 IP 配置

测试联通性（ping 192.168.2.254）正常。

2. 公司分支配置基础 IP 地址。

R1 配置如下所示。

R1(config)#interface e0/1

R1(config-if)#no shutdown

R1(config-if)#ip add 172.16.1.254 255.255.255.0

R1(config-if)#interface e0/2

R1(config-if)#no shutdown

R1(config-if)#ip add 100.1.1.2 255.255.255.0

PC1 配置如下所示。

PC1(config)#no ip routing

PC1(config)#ip default-gateway 172.16.1.254

PC1(config)#interface e0/0

PC1(config-if)#no shutdown

PC1(config-if)#ip add 172.16.1.1 255.255.255.0

3. 安装 JRE，给 PIX 上传 asdm 软件镜像到 PIX，安装 asdm。（具体步骤见前面）此处略。

4. 配置 PIX 的 ASDM 配置，如下所示。

PIX1(config)# http server enable

PIX1(config)# http 192.168.2.0 255.255.255.0 inside

PIX1(config)# username test password test privilege 15

PIX1(config)# asdm image flash:/asdm-613.bin

5. 配置 PIX 的 NAT，依次勾选"Configuration"→"Firewall"→"NAT Rules"→"Add Dynamic NAT Rules"，如图 8-29 所示。

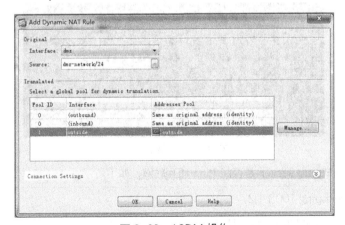

图 8-29　ASDM 操作

6. Manage—Add—Interface：选择"outside"→勾选"Port Address Translation（PAT）"→单击"OK"按钮，如图 8-30 所示。

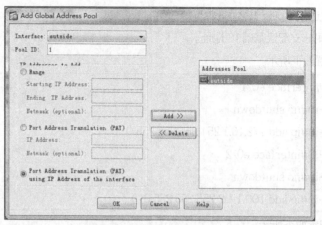

图 8-30　ASDM 操作 2

7. 选择刚刚添加的池，单击"OK"按钮，PAT 就创建配置好了，如图 8-31 所示。

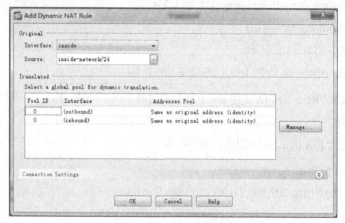

图 8-31　ASDM 操作 3

8. 配置 dmz 区域的 PAT 方法同上，单击"OK"按钮，最后 Apply（应用）。
9. 测试真机是否能 Ping 通互联网，能 Ping 通，说明 NAT 正常，如图 8-32 所示。

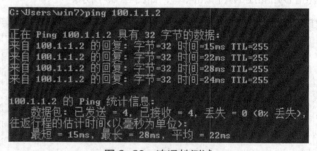

图 8-32　连通性测试

10. 测试 http server 是否能 Ping 通互联网，如下所示。

http_server#ping 100.1.1.2

Type escape sequence to abort.

Sending 5, 100-byte ICMP Echos to 100.1.1.2, timeout is 2 seconds：

!!!!!

Success rate is 100 percent (5/5), round-trip min/avg/max = 20/34/60 ms

11. 左分支的 R1 上配置 NAT，如下所示。

```
R1(config)#access-list 1 permit 172.16.1.0 0.0.0.255
R1(config)#interface e0/1
R1(config-if)#ip nat inside
R1(config-if)#exit
R1(config)#interface e0/2
R1(config-if)#ip nat outside
R1(config-if)#exit
R1(config)#ip nat inside source list 1 interface e0/2 overload
```

12. 测试 PC1 与互联网的连通性，如下所示。

```
PC1#ping 100.1.1.1

Type escape sequence to abort.
Sending 5, 100-byte ICMP Echos to 100.1.1.1, timeout is 2 seconds:
!!!!!
Success rate is 100 percent (5/5), round-trip min/avg/max = 20/36/40 ms
PC1#ping 100.1.1.2

Type escape sequence to abort.
Sending 5, 100-byte ICMP Echos to 100.1.1.2, timeout is 2 seconds:
!!!!!
Success rate is 100 percent (5/5), round-trip min/avg/max = 20/31/76 ms
```

13. 用静态 NAT 实现 PC1 通过端口 8008 访问 80，由于 PC1 用的路由器不好测试，故可以改成通过端口 8008 访问 23，这样我们就可以用 telnet 测试了。

14. 在 NAT Rules 中配置 "Add Static NAT Rule"，如图 8-33 所示。

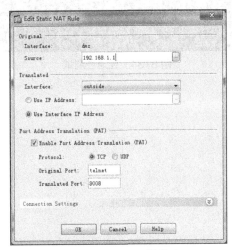

图 8-33　ASDM 操作 4

15. 在 http server 上配置 telnet，如下所示。

http_server(config)#line vty 0 4
http_server(config-line)#no login
http_server(config-line)#privilege level 15

16. 配置一条 acl 放行 telnet 流量（这里为了方便，放行了所有），如下所示。

PIX1(config)# access-list telnet permit ip any any
PIX1(config)# access-group telnet in interface outside

17. 在 PC1 上测试，看端口是否映射成功，如下所示。

PC1#telnet 100.1.1.1 8008
Trying 100.1.1.1, 8008 ... Open

http_server#

可以看到当我们 telnet 100.1.1.1 8008 的时候，防火墙会根据它的 NAT 转换表，自动将其转换成 192.168.1.1 的 23 号端口。

18. 由于默认情况下 inside 访问 dmz 会转换成 outside 接口地址去访问 dmz，故需要加一条 nat 0 拒绝 inside 访问 dmz 转换 NAT 地址。

19. 测试 inside 的 Loopback 口访问 dmz，如下所示。

PIX1(config)#access-list deny-inside extended permit ip 192.168.2.0 255.255.255.0 192.168.1.0 255.255.255.0

20. 测试真机是否能 Ping 通互联网，能 Ping 通，说明 NAT 正常，如图 8-34 所示。

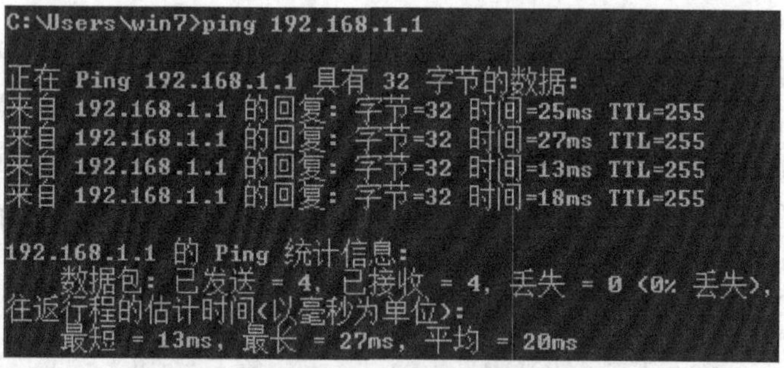

图 8-34　连通性测试

实验成功。

8.3 防火墙的高级应用

实验目的：
1. 掌握 VPN 的基本原理。
2. 掌握在防火墙上通过 ASDM 部署 VPN。

实验拓扑：

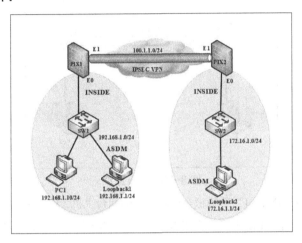

实验随手记：

_____

_____

_____

_____

_____

_____

实验说明：

1. PIX1 和 PIX2 分别为某企业总部和分支的边缘防火墙，通过部署 VPN 技术实现总部和分支能互相通信，且对它们之间通信的数据进行加密。

2. PC1 用 C3640 模拟，交换机用 GNS3 自带的以太网交换机，把真机的 Loopback1 和 Loopback2 桥接到 GNS3 中分别模拟总部和分支的 PC。

实验原理：

1. VPN 概述

Virtual Private Network 即虚拟私有网络，可以在网络中建立逻辑加密隧道，实现数据安全加密，如图 8-35 所示。网络技术中有非常多的 VPN 加密技术，最常见的有 IPsec VPN 和 SSL VPN，本书中主要介绍 IPsec VPN。IPsec VPN 可以实现数据加密性、数据完整性校验、数据不可否认性等功能。IPsec VPN 在企业网、校园网、政务网等应用广泛，它可以在边缘路由器、防火墙或者专用的 VPN 网关产品上进行部署。在思科的解决方案中，根据网络拓扑的需求，例如地址是静态的还是动态的、拓扑是点对点还是多点、数据包是否支持组播广播等可以把 VPN 的解决方案分为 L2L VPN、GRE OVER IPsec、DMVPN、EzVPN 等。本书部署的是最常用的 L2L VPN 解决方案。

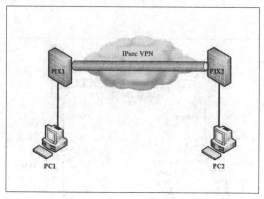

图 8-35　VPN 原理

图 8-35 中，PC1 和 PC2 模拟企业网络中的两台主机，PIX1 和 PIX2 为企业边缘防火墙，在 PIX1 和 PIX2 上建立了 VPN，PC1 和 PC2 通过 VPN 实现数据的安全传输。

2. VPN 原理

IPsec VPN 由 3 个组件组成，分别是 IKE、ESP、AH 协议。3 个组件有不同的功能，IKE（Internet Key Exchanger，互联网密钥交换）负责加密隧道的建立，ESP（Encapsulation Security Payload，封装安全净荷）负责数据的加密和完整性校验，AH（Authentication Header，认证头部）负责数据的完整性校验。IPsec VPN 是一个加密系统，涉及对称或不对称加密算法、哈希算法、身份认证等。以下为 IPsec VPN 隧道的建立过程，如图 8-36 所示。

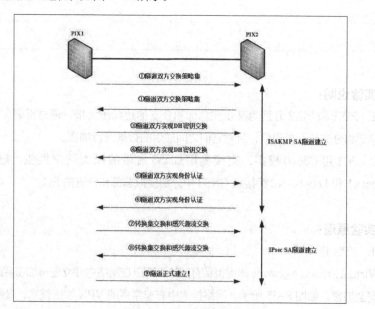

图 8-36　VPN 隧道建立

IPsec VPN 建立过程总共分为两个阶段，即 IKE 阶段一和 IKE 阶段二。其中，阶段一用于建立 ISAKMP SA 隧道，此隧道不是用于加密用户数据，而是加密后续的隧道协商数据包；阶段二用于建立 IPsec Sa 隧道，此隧道是为了加密后续的用户数据流。当 VPN 建立完成之后，后续便进入数据加密过程，数据加密由 ESP 协议来执行，IPsec VPN 对网络层及以上数据进行加密，可以分为两种加密模式，即传输模式（见图 8-37）和隧道模式（见图 8-38）。

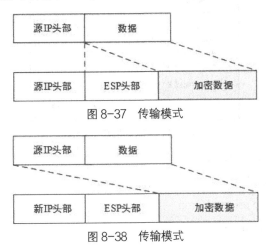

图 8-37　传输模式

图 8-38　传输模式

实验步骤：

1. 公司总部配置基础 IP 地址。

PIX1 上的配置如下所示。

```
PIX1(config)#fixup protocol icmp
PIX1(config)# interface e0
PIX1(config-if)# no shutdown
PIX1(config-if)# ip add 192.168.1.254 255.255.255.0
PIX1(config-if)# nameif inside
INFO: Security level for "inside" set to 100 by default.
PIX1(config)# interface e1
PIX1(config-if)# no shutdown
PIX1(config-if)# ip add 100.1.1.1 255.255.255.0
PIX1(config-if)# nameif outside
INFO: Security level for "outside" set to 0 by default.
```

　PC1 上的配置如下所示。

```
PC1(config)#interface f0/0
PC1(config-if)#no shutdown
PC1(config-if)#ip add 192.168.1.10 255.255.255.0
PC1(config)#no ip routing
PC1(config)#ip default-gateway 192.168.1.254
```

Loopback1 的 IP 配置如图 8-39 所示。

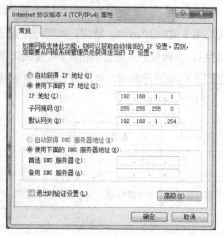

图 8-39　电脑环回口 IP 配置

2. 分支的基础 IP 配置。

PIX2 上的配置如下所示。

```
PIX2(config)#fixup protocol icmp
PIX2(config)# interface e0
PIX2(config-if)# no shutdown
PIX2(config-if)# ip add 172.16.1.254 255.255.255.0
PIX2(config-if)# nameif inside
INFO: Security level for "inside" set to 100 by default.
PIX2(config)# interface e1
PIX2(config-if)# no shutdown
PIX2(config-if)# ip add 100.1.1.2 255.255.255.0
PIX2(config-if)# nameif outside
INFO: Security level for "outside" set to 0 by default.
```

Loopback2 的 IP 配置如图 8-40 所示。

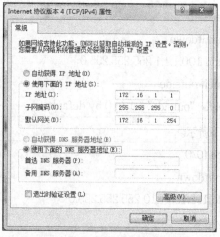

图 8-40　电脑环回口 IP 配置

3. 用 ASDM 为总部和分支配置 NAT（参考前一个实验），此处步骤略。

4. 在 PIX1 上部署 L2L VPN，单击"Wizards"，选择"IPsec VPN Wizard…"，单击"Next"按钮，如图 8-41 所示。

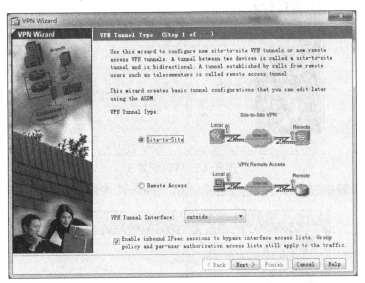

图 8-41　VPN 配置

5. 在"Peer IP Address"选项中填写对方防火墙的公网 IP，在"Pre-Shared Key"选项中填写 cisco，其他默认。

6. 一直单击"Next"按钮到第五步，设置需要加密的流量后，单击"Finish"按钮，如图 8-42 所示。

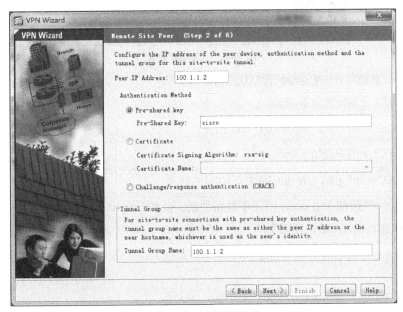

图 8-42　VPN 配置 2

7. 在 PIX2 上部署 L2L VPN，方法同上，如图 8-43 所示。

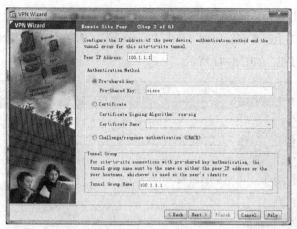

图 8-43 VPN 配置 3

8. 设置分支到总部的加密流量后，单击"Finish"按钮，如图 8-44 所示。

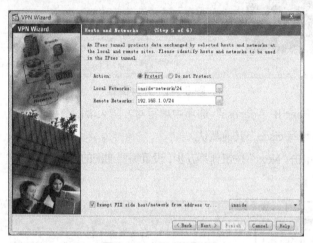

图 8-44 VPN 配置 4

9. 总部和分支的边缘配置默认路由。

PIX1 上的配置如下所示。

PIX1(config)# route outside 0 0 100.1.1.2

PIX2 上的配置如下所示。

PIX2(config)# route outside 0 0 100.1.1.1

10. 禁用网卡 Loopback1（不禁用的话，会和 Loopback2 发生冲突），在 R1 上测试到 Loopback2 的连通性。

R1#ping 172.16.1.1

Type escape sequence to abort.

Sending 5, 100-byte ICMP Echos to 172.16.1.1, timeout is 2 seconds:

.!!!!

Success rate is 80 percent (4/5), round-trip min/avg/max = 76/135/164 ms

通信成功，现在我们已经建立了分支和总部的加密通道。

第 9 章　企业项目实战

本章通过两个企业项目案例来进行实战部署，通过对本章项目案例的部署，可以掌握目前大部分中小型企业网络的架构。

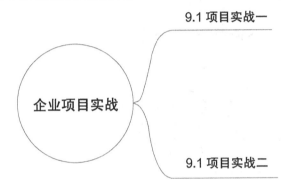

9.1 项目实战一

实验目的：

1. 掌握中小型企业网络的基本部署。
2. 熟悉中小型企业网络的部署流程、排错思路等。

实验拓扑：

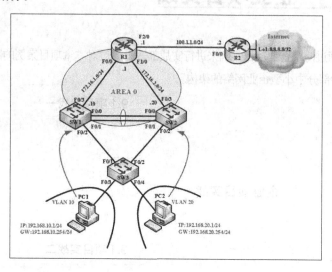

实验随手记：

_____
_____
_____
_____
_____
_____

实验要求：

一、设备管理

1. 依据图中拓扑，为不同设备定义主机名。
2. 全局关闭域名解析。
3. 在 Console 和 VTY 线路下关闭线路超时并开启输出同步。
4. 为实现安全登录，要求创建本地用户名 PingingLab，密码 CCIE，并将其调用到 Console 和 VTY 线路下；要求设置特权密码 CISCO，并要求加密存储。
5. 所有交换机管理 VLAN 为 VLAN1，所在网段为 192.168.1.0/24，其中 SW1 的管理 IP 为 192.168.1.1/24，SW2 为 192.168.1.2/24，SW3 为 192.168.1.3/24，要求能够实现远程管理。

二、交换技术

1. Trunk 技术，具体如下。

①所有交换机之间强制启用 Trunk，并采用 802.1Q 进行封装。

②全局 native vlan 定义为 vlan 10。

③要求 Trunk 上只允许 VLAN1、10、20 的数据通过。

2. VTP 技术，具体如下。

①SW1 为 Server，其他交换机为 Client。

②VTP 管理域为 PingingLab，密码为 cisco。

③全局开启 VTP 修剪。

④在 SW1 上创建 VLAN10/20，并要求全局同步。

⑤将不同用户接口放入相应的 VLAN 中。

3. STP 技术，具体如下。

①部署 PVST，要求 SW1 为 VLAN10 的 Root，VLAN20 的 Secondary；SW2 为 VLAN20 的 Root，VLAN10 的 Secondary，实现负载均衡。

②开启 Portfast，加速用户接入网络接口。

③开启 Uplinkfast，加速直连链路收敛。

④开启 Backbonefast，加速骨干链路收敛。

4. L3 Swithing 技术，具体如下。

①SW1 作为 VLAN10 的主网关、VLAN20 的备网关，其中 VLAN10 地址为 192.168.10.254/24，VLAN20 地址为 192.168.20.253/24；SW2 作为 VLAN20 的主网关、VLAN10 的备网关，其中 VLAN10 地址为 192.168.10.253/24，VLAN20 地址为 192.168.20.254/24。

②在 SW1 和 SW2 上同时部署 DHCP 服务，方便不同 VLAN 的主机接入网络，其中主 DNS 为 8.8.8.8，备用 DNS 为 114.114.114.114。

③在三层交换机上开启三层路由功能，并要求 VLAN 间主机能够相互通信。

5. Etherchannel 技术。为实现链路冗余并提供网络带宽，要求在汇聚层交换机之间部署 L2 Etherchannel 技术。

6. Port-Security 技术，具体如下。

①为实现用户接入安全，要求在所有用户接入接口启用端口安全技术。

②开启地址学习，并定义最大 MAC 数为 1。

③定义用户违反规则为 shutdown 模式，并要求在 30s 后自动恢复。

三、路由技术

1. 在三层交换机 SW1、SW2 和 R1 上部署动态路由协议 OSPF，并通告到骨干区域中。

2. 在边缘路由器 R1 上部署默认路由，用于访问互联网。

四、安全策略

1. 要求只允许管理员地址 192.168.10.1/24 能够远程访问边缘路由器 R1。

2. 为实现内网主机访问互联网，要求部署 PAT 技术。

9.2 项目实战二

实验目的：

1. 掌握中小型企业网络的基本部署。
2. 熟悉中小型企业网络的部署流程、排错思路等。

实验拓扑：

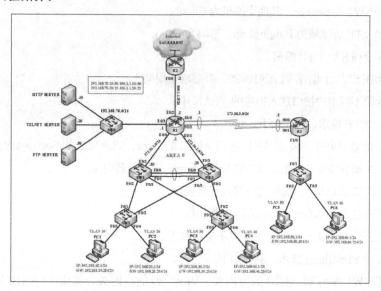

实验随手记：

_____
_____
_____
_____
_____

实验要求：

一、设备管理

1. 依据图中拓扑，为全网设备定义主机名并配置 IP 地址。
2. 全网设备关闭域名解析。
3. 全网设备在 Console 和 VTY 线路下关闭线路超时并开启输出同步。
4. 为实现安全登录，要求在全网设备创建本地用户名 PingingLab，密码 CCIE，并将其调用到 Console 和 VTY 线路下；要求设置特权密码 CISCO，并要求加密存储。
5. 总部所有交换机管理 VLAN 为 VLAN1，所在网段为 192.168.1.0/24，其中

SW1 的管理 IP 为 192.168.1.1/24，SW2 为 192.168.1.2/24，SW3 为 192.168.1.3/24，SW4 为 192.168.1.4/24；分支交换机管理 VLAN 为 VLAN50，地址为 192.168.50.253/24；要求所有交换机可以实现远程管理。

二、交换技术

1. Trunk 技术，具体如下。

①总部所有交换机之间强制启用 Trunk，并采用 802.1Q 进行封装。

②总部所有交换机全局 native vlan 定义为 vlan 10。

③总部所有交换机要求 Trunk 上只允许 VLAN1、10、20、30、40 通过。

2. VTP&VLAN 技术，具体如下。

①总部 SW1 和 SW2 均为 Server，其他交换机为 Client。

②总部 VTP 管理域为 PingingLab，密码为 cisco。

③总部全局开启 VTP 修剪。

④在 SW1 上创建 VLAN10/20/30/40，并要求全局同步。

⑤在分支交换机本地创建 VLAN50 和 VLAN60。

⑥将不同用户接口放入相应的 VLAN 中。

3. STP 技术，具体如下。

①部署 PVST，要求 SW1 为 VLAN10/30 的 Root，VLAN20/40 的 Secondary；SW2 为 VLAN20/40 的 Root，VLAN10/40 的 Secondary，实现负载均衡。

②开启 Portfast，加速用户接入网络接口。

③开启 Uplinkfast，加速直连链路收敛。

④开启 Backbonefast，加速骨干链路收敛。

4. L3 Swithing 技术 & 单臂路由，具体如下。

① SW1 作为 VLAN10/30 的主网关、VLAN20/40 的备网关。

SW2 作为 VLAN20/40 的主网关、VLAN10/30 的备网关，网关地址具体如下。

VLAN10：主 192.168.10.254/24，备 192.168.10.253/24

VLAN20：主 192.168.20.254/24，备 192.168.20.253/24

VLAN30：主 192.168.30.254/24，备 192.168.30.253/24

VLAN40：主 192.168.40.254/24，备 192.168.40.253/24

②在 SW1、SW2、R3 上同时部署 DHCP 服务，方便不同 VLAN 的主机接入网络，其中主 DNS 为 8.8.8.8，备用 DNS 为 114.114.114.114。

③在三层交换机上开启三层路由功能，并要求 VLAN 间主机能够相互通信。

④分支网络要求部署单臂路由，实现 VLAN 间通信。

5. Etherchannel 技术。为实现链路冗余并提供网络带宽，要求在汇聚层交换机之间部署 L2 Etherchannel 技术。

6. Port-Security 技术，具体如下。

①为实现用户接入安全，要求在所有用户接入接口启用端口安全技术。

②开启地址学习，并定义最大 MAC 数为 1。

③定义用户违反规则为 shutdown 模式，并要求在 30s 后自动恢复。

三、路由技术

1. 在全网所有三层设备 SW1、SW2、R1、R3 上部署动态路由协议 OSPF，并通告到骨干区域中。

2. 内网三层设备 R3/SW1/SW2 部署默认路由指向边缘路由器 R1。

3. 在边缘路由器 R1 上部署默认路由，用于访问互联网。

四、安全策略

1. 要求只允许管理员地址 192.168.10.1/24 远程访问边缘路由器 R1 和 R2。

2. 为实现内网主机访问互联网，要求部署 PAT 技术。并且要求 VLAN60 无法访问互联网，VLAN50 只能在周末时段访问互联网，其他 VLAN 正常访问。

3. 为实现互联网主机访问本地服务器，要求部署静态 NAT 技术，将本地的 HTTP 和 Telnet 服务提供出去。

4. 为实现内网互访安全，要求其他 VLAN 无法访问 VLAN60，其他 VLAN 可以相互访问。

5. 为防止设备故障导致配置丢失，要求备份全网设备配置文件，将所有配置文件保存到服务集群中的 TFTP Server。

第 10 章 综合测试

本章主要针对之前所学过的所有知识点做一个整体的理论到实验的测试过程，作为对整本书知识点的反馈和验收。

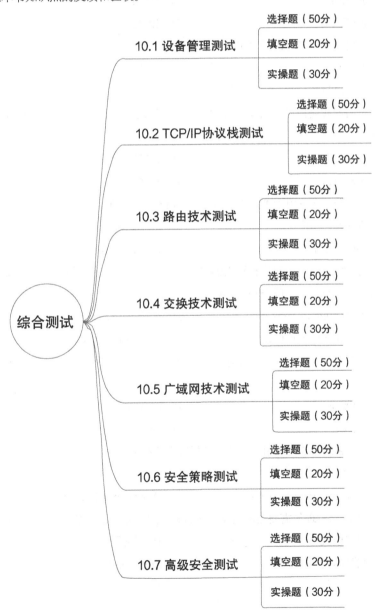

第 10 章 综合测试

10.1 设备管理测试

一、选择题（本大题共 10 小题，每小题 5 分，共 50 分）

1. CISCO IOS 的操作模式包括以下哪几种？（　　）【多选】
   A. 用户模式　　　　　　　　　　B. 特权模式
   C. 配置模式　　　　　　　　　　D. 保护模式

2. 以下哪些命令可以用于保存配置文件？（　　）【多选】
   A. write　　　　　　　　　　　B. copy run start
   C. copy start run　　　　　　　D. 以上都是

3. 思科路由器的哪个部件用于存储配置寄存器？（　　）
   A. FLASH　　　　　　　　　　　B. RAM
   C. NVRAM　　　　　　　　　　　D. ROM

4. 思科交换机与路由器对比，少了哪个部件？（　　）
   A. FLASH　　　　　　　　　　　B. RAM
   C. NVRAM　　　　　　　　　　　D. ROM

5. 路由器和交换机在启动设备的过程中，默认在哪个部件启动网际操作系统？（　　）
   A. FLASH　　　　　　　　　　　B. RAM
   C. NVRAM　　　　　　　　　　　D. ROM

6. 以下哪条命令用于进入路由器的控制口？（　　）
   A. line vty 0 15　　　　　　　B. int fastethernet 0/1
   C. line console 0　　　　　　　D. 以上都不是

7. 当网络管理员在对思科路由器进行操作时，如果设备的日志输出信息对操作造成干扰，可以通过以下哪条命令来解决？（　　）
   A. no ip domain-lookup　　　　B. enable password cisco
   C. logging synchronous　　　　D. exec-timeout 0 0

8. 当网络管理员想修改接口的双工模式时，可以采用以下哪条命令来解决？（　　）
   A. shutdown　　　　　　　　　　B. no shutdown
   C. duplex　　　　　　　　　　　D. speed

9. 以下哪条命令可以用于查看内存配置文件？（　　）
   A. show run　　　　　　　　　　B. show start
   C. show ip interface brief　　D. show cdp neighbor

10. 若网络管理员需要为路由器破解密码，需要进入哪种模式？（　　）
    A. 用户模式　　　　　　　　　　B. 特权模式
    C. 配置模式　　　　　　　　　　D. Rommon 模式

二、填空题（本大题共 4 小题，每小题 5 分，共 20 分）

1. 在配置模式下配置_____命令可以为路由器关闭域名解析。

2. 为了保证密码安全存储，开启全局加密服务，可以在配置模式下配置_____来实现。

3. 通过_____命令可以将路由器本地保存的配置发送到 TFTP 服务器。

4. _____命令可以修改寄存值，使得路由器在 Rommon 模式下实现密码破解。

三、实操题（本大题共 3 小题，每小题 10 分，共 30 分）

1. 要求通过 GNS3 搭建如下拓扑。（10 分）

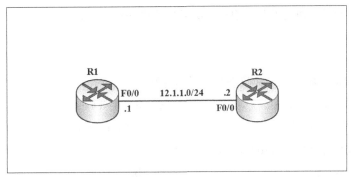

2. 根据以下需求配置 R1 和 R2，如下所示。（10 分）

①定义主机名。

②关闭域名解析。

③关闭线路超时并开启输出同步。

④开启接口并根据拓扑配置 IP 地址，保证直连连通。

⑤配置 console 登录密码为 CISCO，配置特权密码为 CCNA，特权密码采用加密方式。

⑥保存配置。

3. 在 R2 上开启 FTP 服务，将 R1 保存配置传送到 R2 上。（10 分）

10.2 TCP/IP 协议栈测试

一、选择题（本大题共 10 小题，每小题 5 分，共 50 分）

1. 下面哪个协议可以实现可靠传输功能？（　　）
   A. IP　　　　　　　　　　　B. UDP
   C. ARP　　　　　　　　　　D. TCP

2. 下面哪条命令可以实现链路连通性测试？（　　）
   A. ping　　　　　　　　　　B. ifconfig
   C. netstate　　　　　　　　D. netsh

3. 下面哪个地址为 A 类地址？（　　）
   A. 127.1.1.1　　　　　　　　B. 10.1.1.1
   C. 192.168.1.1　　　　　　　D. 225.100.100.100

4. 下面哪个协议可用于实现 IP 到 MAC 地址的映射？（　　）
   A. RARP　　　　　　　　　　B. ARP
   C. ICMP　　　　　　　　　　D. UDP

5. DHCP 客户端局域网发送哪个数据包用于发现 DHCP 服务器？（　　）
   A. DHCP Request　　　　　　B. DHCP Release
   C. DHCP ACK　　　　　　　　D. DHCP Discover

6. 本地环回测试的 IP 网段是（　　）。
   A. 164.0.0.0　　　　　　　　B. 130.0.0.0
   C. 200.0.0.0　　　　　　　　D. 127.0.0.0

7. DHCP 客户端是使用地址（　　）来申请一个新的 IP 地址的。
   A. 0.0.0.0　　　　　　　　　B. 10.0.0.1
   C. 127.0.0.1　　　　　　　　D. 255.255.255.255

8. 相对于 OSI 模型，TCP/IP 协议栈少了哪些层？（　　）【多选】
   A. 应用层　　　　　　　　　B. 会话层
   C. 表示层　　　　　　　　　D. 网络层

9. 下面哪些协议会发送广播包？（　　）【多选】
   A. ARP　　　　　　　　　　B. TCP
   C. OSPF　　　　　　　　　　D. DHCP

10. DHCP 通讯的端口有哪些？（　　）【多选】
    A. 67　　　　　　　　　　　B. 68
    C. 69　　　　　　　　　　　D. 70

二、填空题（本大题共 4 小题，每小题 5 分，共 20 分）

1. OSI 七层模型从下到上分别为物理层、_____、网络层、_____、会话层、表示层和_____。

2. OSI 的_____层提供端到端的可靠传输服务。

3. DHCP客户端获取地址过程涉及DHCP发现、_____、DHCP请求、_____。

4. 在以太网中ARP报文分为ARP请求分组和ARP回应分组，其中ARP Request采用_____发送，ARP Response采用单播包发送。

三、实操题（本大题共3小题，其中第1题5分，第2题5分，第3题20分，共30分）

1. 要求通过GNS3搭建如下拓扑。（5分）

```
        R1                          R2
       F0/0    12.1.1.0/24   .2
               .1            F0/0
```

2. 配置R1和R2路由器，并包括以下需求。（5分）

①定义主机名。

②打开接口，并为接口配置IP地址。

请写下R1和R2的配置：

R1=>

R2=>

3. 在GNS3上开启Wireshark抓包，抓取R1和R2数据分组，在R1上Ping R2，抓取ARP、IP、ICMP三种协议报文，要求将此三种分组视图截图。（20分）

ARP分组视图 =>

IP 分组视图 =>

ICMP 分组视图 =>

10.3 路由技术测试

一、选择题（本大题共 10 小题，每小题 5 分，共 50 分）

1. 下面哪个算法是 RIP 的路由算法？（   ）
   A. TCP   B. BGP
   C. SPF   D. Bellman

2. 下面哪些路由协议属于距离矢量协议？（   ）【多选】
   A. RIPv1   B. RIPv2
   C. EIGRP   D. OSPF

3. 下面哪些路由协议属于无类路由协议？（   ）【多选】
   A. RIPv1   B. RIPv2
   C. EIGRP   D. OSPF

4. RIPv2 的组播更新地址为（   ）。
   A. 224.0.0.5   B. 224.0.0.6
   C. 224.0.0.7   D. 224.0.0.9

5. 若 RIPv2 本地有四个子网，分别为 192.168.4.0/24、192.168.5.0/24、192.168.6.0/24、192.168.7.0/24，则可以汇总为（   ）。
   A. 192.168.0.0/22   B. 192.168.0.0/21
   C. 192.168.4.0/21   D. 192.168.4.0/22

6. 默认路由的写法为 ip route（   ）+ 出接口 / 下一跳 IP。
   A. 0.0.0.0 255.255.255.255   B. 255.255.255.255 0.0.0.0
   C. 255.255.255.255 255.255.255.255 D. 0.0.0.0 0.0.0.0

7. OSPF 路由协议采用哪种算法进行路由计算？（   ）
   A. TCP   B. BGP
   C. SPF   D. Bellman

8. OSPF 的区域 0 代表了（   ）。
   A. 初始区域   B. 末节区域
   C. 常规区域   D. 骨干区域

9. 相比距离矢量协议，链路状态协议具备以下哪些特征？（   ）【多选】
   A. 基于跳数进行路由选择   B. 以自己为中心，在本地进行路由计算
   C. 路由器之间交互路由条目   D. 路由器之间交互链路状态信息

10. 以下哪些路由协议可以进行可靠路由的更新？（   ）【多选】
    A. RIPv1   B. RIPv2
    C. EIGRP   D. OSPF

二、填空题（本大题共 4 小题，每小题 5 分，共 20 分）

1. RIP 是基于 UDP 协议，端口号为_____的路由协议。

2. RIP 的防环机制包括水平分割_____、_____、_____、抑制计时器。

3. OSPF 和 EIGRP 是可靠更新协议，都是基于 OSI 的_____层。
4. 在 OSPF 协议中，_____分组是用于建立和维持邻居关系的。

三、实操题（本大题共 2 小题，每小题 15 分，共 30 分）

1. 静态路由。（15 分）

（1）要求通过 GNS3 搭建如下拓扑。（5 分）

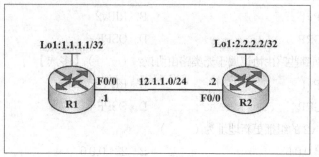

（2）在 R1 和 R2 上开启接口并配置 IP 地址，要求直连相互 Ping 通。（5 分）

（3）部署静态路由，并要求相互 Ping 通对方环回接口。（5 分）

2. OSPF 路由协议。（15 分）

（1）要求通过 GNS3 搭建如下拓扑。（5 分）

（2）为图中 R1、R2、R3 开启接口并配置 IP 地址，并保证直连连通（5 分）

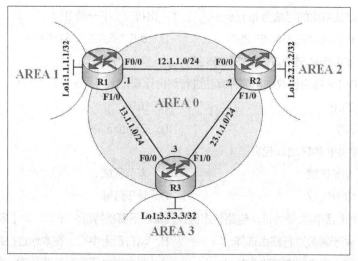

（3）在 R1、R2、R3 上部署 OSPF 协议，实现区域间路由通信，要求路由器能够 Ping 通对方环回接口。（5 分）

10.4 交换技术测试

一、选择题（本大题共 10 小题，每小题 5 分，共 50 分）

1. 交换机具备以下哪些功能？（   ）【多选】
   A. 数据帧转发   B. 地址学习
   C. 防止环路   D. IP 包过滤

2. 以下哪些技术可以用于防止环路？（   ）【多选】
   A. STP   B. VTP
   C. PVST   D. VLAN

3. VTP 协议用于实现 VLAN 统一管理，具备以下哪些模式？（   ）【多选】
   A. Server 模式   B. Client 模式
   C. Transparent 模式   D. 以上都不是

4. 以太网交换机的一个端口在接收到数据侦时，如果没有在 MAC 地址表中查找到目的 MAC 地址，通常如何处理？（   ）
   A. 把以太网侦复制到所有端口
   B. 把以太网侦单点传送到特定端口
   C. 把以太网侦发送到除本端口以外的所有端口
   D. 丢弃该侦

5. 启用 STP 的缺省情况下，以下哪个交换机将被选为根桥？（   ）
   A. 拥有最小 MAC 地址的交换机
   B. 拥有最大 MAC 地址的交换机
   C. 端口优先级数值最高的交换机
   D. 端口优先级数值最低的交换机

6. 下列哪种安全技术可以实现端口安全？（   ）
   A. Port-Security   B. DAI
   C. IPSG   D. DHCP Snooping

7. 若网络管理员需要提高网络带宽并实现链路冗余，可以通过以下哪种技术来实现？（   ）
   A. Etherchannel   B. Trunk
   C. VLAN   D. DHCP

8. 端口安全 Port-Security 的违规模式有哪几种？（   ）【多选】
   A. shutdown   B. restrict
   C. protect   D. blocking

9. 目前局域网架构采用分层架构，它主要包括哪些层？（   ）【多选】
   A. 核心层   B. 汇聚层
   C. 接入层   D. 安全层

10. 以下哪种技术可以用于承载不同 VLAN 的流量并且为数据打上标签？（   ）
    A. VLAN   B. VTP
    C. Trunk   D. STP

二、填空题（本大题共 4 小题，每小题 5 分，共 20 分）

1. STP 的端口状态机包括 Blocking、_____、Learning、_____。
2. _____协议可以防环也可以实现流量负载均衡。
3. VTP 的_____模式可以防止新加入的交换机干扰原有网络的 vlan 信息。
4. 通过生成树的加速特性 portfast、_____、_____可以有效提高生成树的收敛时间。

三、实操题（本大题共 6 小题，每小题 5 分，共 30 分）

1. 要求通过 GNS3 搭建如下拓扑。（5 分）

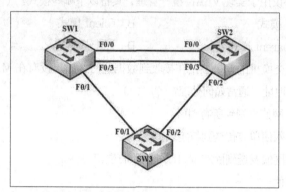

2. 为 SW1、SW2、SW3 配置主机名，开启接口，并强制为 Trunk。（5 分）
3. 部署 VTP 协议，要求 VTP 域为 CISCO，密码为 CCNA，SW1 为服务模式，SW2 和 SW3 为客户模式，要求在 SW1 上创建 VLAN10、20，SW2 和 SW3 能够通过 VTP 学习到。（5 分）
4. 部署 PVST 协议，要求 SW1 为 VLAN10 的根，VLAN20 的备根，SW2 为 VLAN20 的根，VLAN10 的备根。（5 分）
5. 在 SW3 上开启 uplinkfast 特性，在所有交换机上开启 backbonefast。（5 分）
6. 在 SW1 和 SW2 上开启 Etherchannel，对接口进行强制捆绑。（5 分）

10.5 广域网技术测试

一、选择题（本大题共 10 小题，每小题 5 分，共 50 分）

1. 以下哪些协议是点对点链路上使用的？（   ）【多选】
   A. PPP   B. HDLC
   C. Frame-Relay   D. Ethernet

2. 以下哪种广域网协议可以实现安全认证？（   ）
   A. PPP   B. HDLC
   C. Frame-Relay   D. Ethernet

3. PPP 协议有哪些子协议？（   ）【多选】
   A. HDLC   B. LCP
   C. NCP   D. VTP

4. 帧中继技术是一种运行在（   ）环境下的技术。
   A. 多路访问网络   B. 点对点网络
   C. 非广播多路访问网络   D. 以上都不是

5. 以下哪些专业术语属于帧中继技术？（   ）【多选】
   A. PVC   B. DLCI
   C. Inverse arp   D. LCP

6. 当思科路由器和华为路由器要进行点对点连接时，可以采用以下哪种协议？（   ）
   A. HDLC   B. PPP
   C. Frame-Relay   D. Ethernet

7. 以下哪种 PPP 认证可以实现安全加密？（   ）
   A. LCP   B. NCP
   C. PAP   D. CHAP

8. 帧中继环境下，用于实现 IP 到 DLCI 的映射是以下哪种协议？（   ）
   A. ARP   B. RARP
   C. Inverse ARP   D. 以上都不是

9. 在帧中继环境下，路由器用于存储 IP 到 DLCI 的表项称为（   ）。
   A. 帧中继路由表   B. 帧中继转发表
   C. 帧中继映射表   D. 以上都不是

10. 在帧中继环境下，帧中继交换机用于转发数据帧的表项称为（   ）。
    A. 帧中继路由表   B. 帧中继转发表
    C. 帧中继映射表   D. 以上都不是

二、填空题（本大题共 4 小题，每小题 5 分，共 20 分）

1. PPP 具备安全认证功能，包含_____和_____两种认证方式。

2. 帧中继技术中的_____类似以太网中的 MAC 地址。

3. PPP 的_____功能可以有效提高链路带宽和冗余性。
4. 在接口下开启帧中继封装，可以采用命令_____来实现。

三、实操题（本大题共 4 小题，其中第 1 题 5 分，第 2、3 题 10 分，第 4 题 5 分，共 30 分）

1. 要求通过 GNS3 搭建如下拓扑。（5 分）

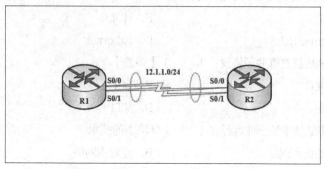

2. 为 R1、R2 定义主机名，开启接口，并部署 PPP 捆绑。（10 分）
3. 在 R1 和 R2 上部署 PPP CHAP 认证，共享密码为 CCNA。（10 分）
4. 在 R1 和 R2 上部署 IP 地址，并要求相互 Ping 通。（5 分）

10.6 安全策略测试

一、选择题（本大题共 10 小题，每小题 5 分，共 50 分）

1. 以下哪种技术可以实现对数据流的过滤？（   ）
   - A. NAT
   - B. OSPF
   - C. ACL
   - D. RIP

2. 标准 ACL 可以实现对以下哪些信息的抓取？（   ）
   - A. 源 IP 地址
   - B. 源端口
   - C. 目的端口
   - D. 目的 IP 地址

3. 拓展 ACL 可以实现对以下哪些信息的抓取？（   ）【多选】
   - A. 源 IP 地址
   - B. 源端口
   - C. 目的端口
   - D. 目的 IP 地址

4. NAT 技术有以下哪些功能？（   ）【多选】
   - A. 地址转换
   - B. 隐藏内部网络
   - C. 保证内网安全
   - D. 以上都不是

5. 以下哪种 NAT 技术可以将内网的服务器映射到互联网上？（   ）
   - A. 动态 NAT
   - B. 端口 NAT
   - C. 静态 NAT
   - D. 以上都不是

6. 若公司网络管理员需要在上班时间禁止办公人员上网，可以采用以下哪种技术来实现？（   ）
   - A. 标准 ACL
   - B. 拓展 ACL
   - C. 编号 ACL
   - D. 时间 ACL

7. 相比编号 ACL，命名 ACL 具备以下哪些特征？（   ）【多选】
   - A. 可删除某条 ACL 语句
   - B. 可插入某条 ACL 语句
   - C. 方便管理，方便记忆
   - D. 以上都不是

8. NAT 技术一般在企业网络的哪个位置进行部署？（   ）
   - A. 接入层交换机
   - B. 汇聚层交换机
   - C. 核心层交换机
   - D. 边缘路由器

9. 以下哪些专业术语属于 NAT 技术的？（   ）【多选】
   - A. 内部本地地址
   - B. 内部全局地址
   - C. 外部本地地址
   - D. 外部全局地址

10. 端口 NAT 和动态 NAT 对比，具备哪些特征？（   ）【多选】
    - A. 地址采用一对一映射
    - B. 更加节约地址空间
    - C. 地址采用多对少映射
    - D. 以上都不是

二、填空题（本大题共 4 小题，每小题 5 分，共 20 分）

1. NAT 可以实现地址转换，它的全称为_____。
2. ACL 根据不同分类方式，可以分为标准 ACL、_____、编号

ACL、_____。

3. 公司内网有200台主机，若公司只申请了2个公有地址，可以采用_____ _____来进行地址转换。

4. 需要创建 NAT 地址池 "NATPOOL"，地址为 100.1.1.1 到 100.1.1.10，写法为_____。

三、实操题（本大题共5小题，其中第4题10分，其他题均5分，共30分）

1. 要求通过 GNS3 搭建如下拓扑。（5分）

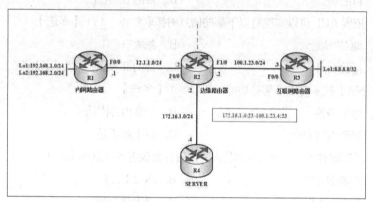

2. 为图中所有路由器定义主机名，开启接口，并配置 IP 地址，保证所有直连网段通信。（5分）

3. 在 R1 和 R4 上部署默认路由到 R2，R2 上部署默认路由到 R3。（5分）

4. 在 R2 上部署端口 NAT 技术（PAT），使得内网网段 192.168.1.0/24 和 192.168.2.0/24 能够通过 R2 出接口地址 100.1.23.2 访问 R3。（10分）

5. 在 R2 上部署静态 NAT 技术，将 R4 的 23 端口映射到公有地址 100.1.23.4 的 23 端口，并通过 R3 进行测试。（5分）

10.7 高级安全测试

一、选择题（本大题共 10 小题，每小题 5 分，共 50 分）

1. 以下哪些网络设备属于安全产品？（　　）
   - A. 路由器
   - B. 交换机
   - C. 防火墙
   - D. 无线 AP

2. 防火墙一般部署在网络的边缘，用于实现（　　）。【多选】
   - A. 攻击防护
   - B. 数据转发
   - C. 地址转换
   - D. 以上都不是

3. 以下哪种技术不是实现防火墙的主流技术？（　　）
   - A. 包过滤技术
   - B. 应用级网关技术
   - C. 代理服务器技术
   - D. NAT 技术

4. （　　）防火墙是在网络的入口对通过的数据包进行选择，只有满足条件的数据包才能通过，否则被抛弃。
   - A. 包过滤
   - B. 应用网关
   - C. 帧过滤
   - D. 代理

5. 为了增强访问网页的安全性，可以采用（　　）协议。
   - A. Telnet
   - B. POP3
   - C. HTTPS
   - D. DNS

6. 包过滤技术可以允许或不允许数据包在网络上传递，它过滤的依据不包括（　　）。
   - A. 数据包的目的地址
   - B. 数据包的源地址
   - C. 数据包的传送协议
   - D. 数据包的具体内容

7. 目前主流防火墙的核心技术是（　　）。
   - A. 简单包过滤技术
   - B. 复合技术
   - C. 应用代理技术
   - D. 状态检测包过滤技术

8. 以下哪些 VPN 技术可以实现安全加密？（　　）【多选】
   - A. IPsec VPN
   - B. MPLS VPN
   - C. SSL VPN
   - D. 以上都是

9. IPsec VPN 可以实现（　　）。
   - A. 入侵防御
   - B. 安全防护
   - C. 数据加密
   - D. 以上都不是

10. IPsec VPN 涉及哪些安全技术？（　　）【多选】
    - A. 对称密钥
    - B. 不对称密钥
    - C. 哈希算法
    - D. 以上都是

二、填空题（本大题共 4 小题，每小题 5 分，共 20 分）

1. 防火墙主要分为包过滤防火墙、_____、_____。

2. 包过滤防火墙是工作在 OSI 层的_____、_____。
3. IPsec VPN 用于加密 OSI 层_____以上的数据。
4. 在部署 VPN 技术时，需要结合_____技术来匹配感兴趣的流量。

三、实操题（30分）

本章节测试题采用 8.1 实验。

附录 术语索引

GNS3，1.1
命名行界面，1.1
图形化界面，1.1
CCNA，1.1
CCNP，1.1
CCIE，1.1
拓扑，1.3
接口模块，1.3
接口标识，1.3
终端登录，1.3
Telnet，1.3
SSH，1.3
Serial，1.3
127.0.0.1，1.3
抓包，1.3
Ping，1.3
VMware，1.4
VirtualBox，1.4
桥接，1.4
帧中继交换机，1.6
DLCI，1.6
PIX & ASA，1.7
序列号 & 激活码，1.7
思科简介，2.1
IOS 背景，2.1
Console 介绍，2.1
日志信息，2.1
常用快捷键，2.1
主机名，2.2
域名解析，2.2

关于时间，2.3
接口分类，2.4
接口状态，2.4
接口地址，2.4
配置保存，2.5
配置备份，2.5
管理方式，2.6
密码安全，2.6
FLASH，2.7
ARP 协议，3.1
数据封装，3.1
广播，3.1
ARP 表，3.1
IP 协议，3.2
面向无连接，3.2
不可靠传输，3.2
QoS，3.2
IP 分片，3.2
ICMP 协议，3.3
类型值 & 代码值，3.3
UDP 协议，3.4
DHCP 协议，3.4
动态 IP，3.4
网关，3.4
TCP 协议，3.5
Telnet 协议，3.5
面向连接，3.5
可靠传输，3.5
滑动窗口，3.5
静态路由，4.1

附录 术语索引

路由表，4.1
环回接口，4.1
静态路由写法，4.1
路由代码，4.1
带源 Ping，4.1
默认路由，4.2
全 0 IP，4.2
浮动路由，4.3
管理距离，4.3
Debug 调试，4.3
动态路由协议，4.4
RIP 协议，4.4
Bellman 算法，4.4
有类和无类，4.4
度量值，4.4
跳数，4.4
路由汇总，4.6
从地址，4.6
EIGRP 协议，4.7
DUAL 算法，4.7
高级距离矢量协议，4.7
触发更新，4.7
组播更新，4.7
增量更新，4.7
自治系统，4.7
自动汇总，4.7
OSPF 协议，4.9
链路状态协议，4.9
SPF 算法，4.9
开销，4.9

进程号，4.9
区域号，4.9
区域，4.10
ABR 路由器，4.10
VLAN，5.1
广播域，5.1
Trunk，5.3
Trunk 封装，5.3
Native VLAN，5.4
Allow VLAN，5.4
DTP 协议，5.5
VTP 协议，5.6
配置版本号，5.6
STP 协议，5.7
BPDU，5.7
BID，5.7
PID，5.7
COST，5.7
单臂路由，5.10
三层交换机，5.11
以太通道，5.13
广域网，6.1
HDLC，6.1
DCE & DTE，6.1
PPP，6.2
明文认证，6.3
密文认证，6.3
Frame-Relay，6.3
帧中继交换机，6.3
ACL，7.1